SpringerBriefs in Economics

SpringerBriefs present concise summaries of cutting-edge research and practical applications across a wide spectrum of fields. Featuring compact volumes of 50 to 125 pages, the series covers a range of content from professional to academic. Typical topics might include:

- A timely report of state-of-the art analytical techniques
- A bridge between new research results, as published in journal articles, and a contextual literature review
- A snapshot of a hot or emerging topic
- An in-depth case study or clinical example
- A presentation of core concepts that students must understand in order to make independent contributions

SpringerBriefs in Economics showcase emerging theory, empirical research, and practical application in microeconomics, macroeconomics, economic policy, public finance, econometrics, regional science, and related fields, from a global author community.

Briefs are characterized by fast, global electronic dissemination, standard publishing contracts, standardized manuscript preparation and formatting guidelines, and expedited production schedules.

More information about this series at http://www.springer.com/series/8876

Carlos Mendez

Convergence Clubs in Labor Productivity and its Proximate Sources

Evidence from Developed and Developing Countries

Springer

Carlos Mendez
Graduate School of International
Development (GSID)
Nagoya University
Nagoya-shi, Japan

ISSN 2191-5504 ISSN 2191-5512 (electronic)
SpringerBriefs in Economics
ISBN 978-981-15-8628-6 ISBN 978-981-15-8629-3 (eBook)
https://doi.org/10.1007/978-981-15-8629-3

This Springer imprint is published by the registered company Springer Nature Singapore Pte Ltd.
The registered company address is: 152 Beach Road, #21-01/04 Gateway East, Singapore 189721, Singapore

To my beloved family.

Preface

Labor productivity is usually pointed out as a central factor to explain the wealth and development of nations. Labor productivity disparities across countries have drastically increased over time. In the twenty-first century, these disparities are at least 40 orders of magnitude.

Testing for economic convergence across countries has been a central issue in the literature of economic growth and development. Initial studies evaluated convergence assuming technological homogeneity across countries. Thus, convergence toward a unique equilibrium is largely expected. More recent contributions, however, have emphasized the role of technological heterogeneity and the existence of multiple convergence clubs.

In this context, this book introduces a club convergence framework to study the cross-country dynamics of labor productivity and its proximate sources: capital accumulation and aggregate efficiency. In particular, recent convergence dynamics of developed as well as developing countries are evaluated through the lens of a non-linear dynamic factor model and a clustering algorithm for panel data. This modern framework allows us to examine key economic phenomena such as technological heterogeneity and multiple equilibria.

Overall, the book provides a succinct review of the recent club convergence literature, a comparative view of developed and developing countries, and a tutorial on how to implement the club convergence framework in the statistical software Stata. These three features could help graduate students and researchers catch up with the latest developments and methodological implementations of the club convergence literature.

The book is organized into seven chapters. Chapter 1 introduces the purpose of the book and outlines the research questions that will be answered in subsequent chapters. Chapter 2 describes a methodological approach to measure labor productivity, capital accumulation, and aggregate efficiency across countries and over time. The chapter empirically illustrates the dynamics of each variable based on the quantiles of the cross-country distribution and their evolution. Chapter 3 presents a modern framework to study convergence dynamics across countries. It includes a selected overview of the club convergence literature and a tutorial on how to implement the framework using the statistical software Stata. Chapter 4 presents the results of the convergence dynamics of labor productivity. Chapters 5 and 6 present the convergence dynamics of capital accumulation and aggregate efficiency, respectively. Finally, Chap. 7 summarizes the main takeaways of the book and points out some emerging research directions.

Nagoya, Japan
July 2020

Carlos Mendez

Acknowledgements

This book was written based on my lecture notes for the Development Macroeconomics course and Economic Development seminar I have been teaching at the Graduate School of International Development (GSID) of Nagoya University. I deeply thank all my students for attending my courses and stimulating discussions about the increasing disparities we observe across countries. Also, I would like to thank all members of the Quantitative Regional and Computational Science lab (QuaRCS-lab) for their research ideas, questions, and content suggestions. Finally, the research assistance and editorial support of Felipe Santos-Marquez have largely improved the content and presentation of this book.

Contents

Chapter 1
Introduction and Overview

Abstract This book introduces a modern club convergence framework to study the cross-country dynamics of labor productivity and its proximate sources: capital accumulation and aggregate efficiency. This chapter starts by outlining the contents of the book in terms of research questions that will be answered in subsequent chapters. Next, it provides a first overview of the data by illustrating the large and increasing productivity differences that observed across countries. It concludes by pointing out the need for a club convergence framework to further improve our understanding of the economic performance of developed and developing countries.

Keywords Convergence · Labor productivity · Physical capital · Human capital · Development · Productivity determinants

1.1 Purpose and Research Questions

At least since the writings of Adam Smith, labor productivity has been pointed out as the central factor to explain the wealth and development of nations. At the time of Adam Smith's writing, labor productivity disparities across countries were reported to be roughly four orders of magnitude. That is, the most productive countries used to be about four times more productive than the least productive countries. Labor productivity disparities across countries have drastically increased over time. In the twenty first century, these disparities are at least forty orders of magnitude. This book studies the evolution of labor productivity disparities in recent years through the lens of a modern framework of economic convergence.

Testing for economic convergence across countries has been a central issue in the literature of economic growth and development (Durlauf et al. 2009; Islam 2003; Johnson and Papageorgiou 2020). Classical contributions, such as those of Baumol (1986) and Barro and Sala-i-Martin (1992), suggest that, conditional on homogeneous technologies, initially poor countries tend to grow faster that initially rich countries. As this process continues, convergence towards a unique equilibrium is

C. Mendez, *Convergence Clubs in Labor Productivity and its Proximate Sources*, SpringerBriefs in Economics, https://doi.org/10.1007/978-981-15-8629-3_1

expected. More recent contributions, however, have emphasized the role of technological heterogeneity and the existence of multiple convergence clubs across countries (Phillips 2007b; Sul 2009; Pittau et al. 2010; Quah 1997)

The main purpose of this book is to provide a brief overview of the convergence framework developed by Phillips and Sul (2007a), Phillips (2007b)), and Sul (2009); and its methodological implementation in the context labor productivity differences across countries. Through the lens of a non-linear dynamic factor model and a clustering algorithm for panel data, this book evaluates the cross-country dynamics of labor productivity and its proximate determinants (capital accumulation and aggregate efficiency) for developed and developing countries, respectively. Specifically, each chapter of this book aims to provide a succinct reference for answering the following questions:

- How large are the labor productivity disparities across developed and developing countries? How are these disparities evolving in recent years? (Current chapter and Chap. 2)
- How to measure labor productivity and its proximate determinants: physical capital, human capital, and aggregate efficiency? What stylized facts can be identified? (Chap. 2)
- How to evaluate the hypothesis that all countries would eventually converge to a common growth path? If all countries do not convergence to a common growth path, how to identify convergence clubs? How to use the statistical software Stata to study club convergence? (Chap. 3)
- In terms of labor productivity dynamics, are developed and developing countries characterized by a unique common growth path or by multiple convergence clubs? (Chap. 4)
- In terms of capital accumulation dynamics, are developed and developing countries characterized by a unique common growth path or by multiple convergence clubs? (Chap. 5)
- In terms of aggregate efficiency dynamics, are developed and developing countries characterized by a unique common growth path or by multiple convergence clubs? (Chap. 6)
- What are some new research directions to study club convergence? (Chap. 7).

1.2 A First Overview of the Data

1.2.1 Labor Productivity Differences Across Countries and Over Time

There are large and increasing gaps in labor productivity across countries. Figures 1.1 and 1.2 illustrate the cross-country spatial distribution of labor productivity in 1990 and 2014, respectively. Labor productivity is measured in a logarithmic scale and

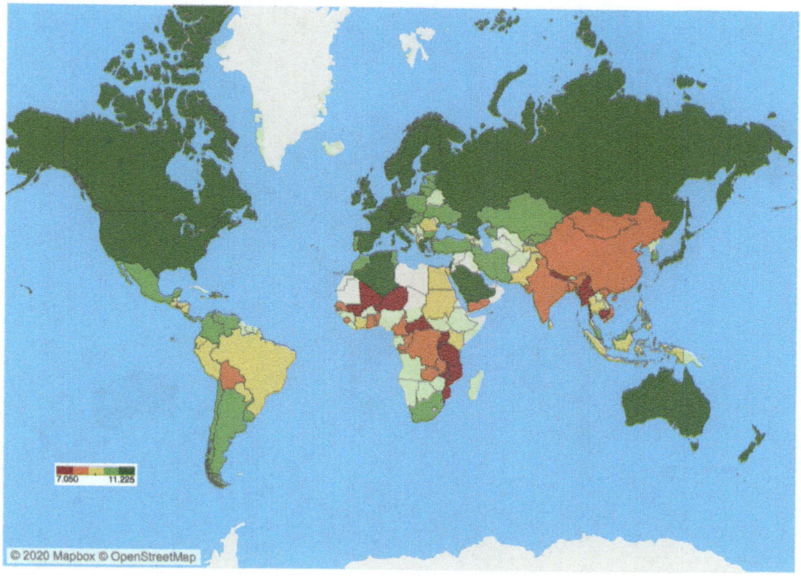

Fig. 1.1 (Log) Differences in labor productivity in 1990

countries are classified into five productivity levels.[1] Geographically, middle and low productivity countries tend to be located near the Equator, while high productivity countries tend to be located in the north and south of the map. On the one hand, as indicated by Fig. 1.2, some countries in Asia have largely improved their labor productivity levels. On the other hand, countries in central Africa show very little progress over the entire 1990–2014 period. This first glimpse of the performance of Asia and Africa illustrates that, across developing countries, convergence in labor productivity has not yet been achieved.

Figure 1.3 shows the entire paths of labor productivity (GDP per worker) for a sample of developed and developing countries over the 1990–2014 period. In this figure, and in the rest of the book, countries are classified into two groups (developed and developing) based on their per-capita income status in 1990. Countries defined by the World Bank as high-income countries in 1990 are labeled as developed countries. All the remaining countries in the sample are labeled as developing countries. As expected, based on this income classification, high (low) income countries tend to have high (low) productivity levels. Moreover, disparities in labor productivity are largely increasing over time.

[1] See Chap. 2 for further details about the measurement of labor productivity and its determinants.

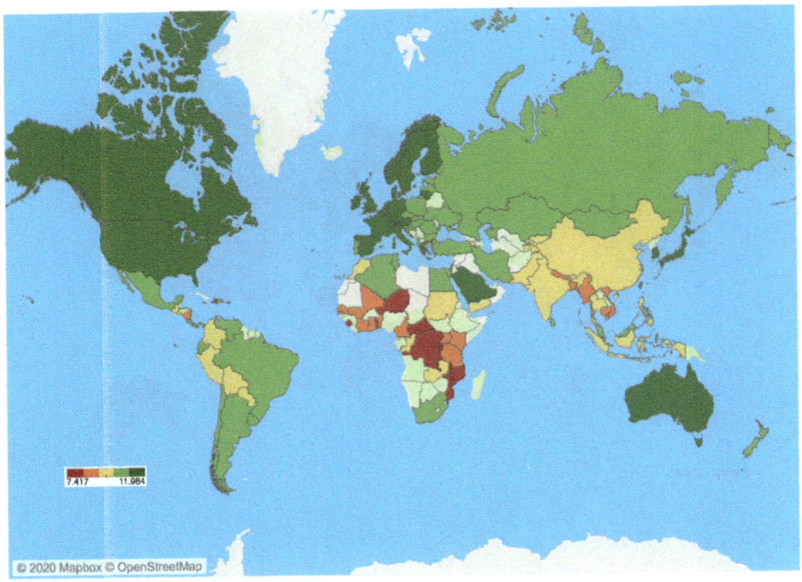

Fig. 1.2 (Log) Differences in labor productivity in 2014

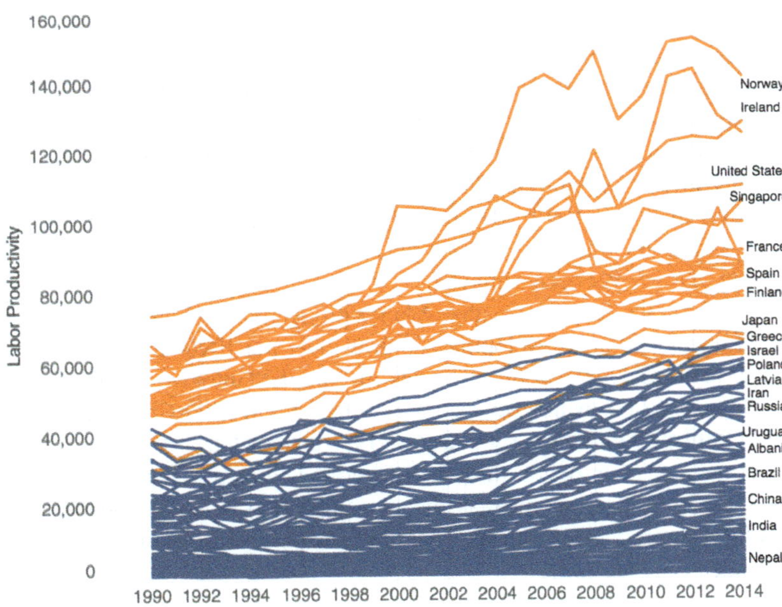

Fig. 1.3 Labor productivity dynamics 1990–2014. *Notes*: Labor productivity is measured as real GDP per worker. In turn, GDP is measured using purchasing power parities (PPPs) data from the 2011 International Comparison Program of the World Bank. *Data source*: All figures of this book and related calculations are largely based on the productivity database of Daude and Fernandez-Arias (2017). See Chap. 2 for further details about the data

1.2.2 Are There Any Signs of Productivity Convergence?

The classical approach to empirically study economic convergence across countries was developed by Baumol (1986) and Barro and Sala-i-Martin (1992). Based on this approach, evidence of productivity convergence would be summarized by the inverse relationship between the growth rate of productivity and its initial level. If such relation holds in the data, it would imply that the productivity level of initially low-productivity countries grows faster than of high-productivity countries. Thus, there is a process of convergence across countries. In the growth econometrics literature (Durlauf et al. 2009), this inverse relationship between the initial level of a variable and its subsequent growth rate is known as beta convergence.

Figure 1.4 shows the strength of this inverse relationship in the context of developed and developing countries. Only among developed countries, low-productivity economies tend to grow faster than high-productivity economies. In contrast, across developing countries, the beta convergence relationship is not statistically significant.

Figure 1.5 provides a much detailed perspective on the long-run paths of labor productivity in developing countries. In this figure, the long-run trend of labor productivity of each country is estimated and then re-scaled by the cross-sectional mean

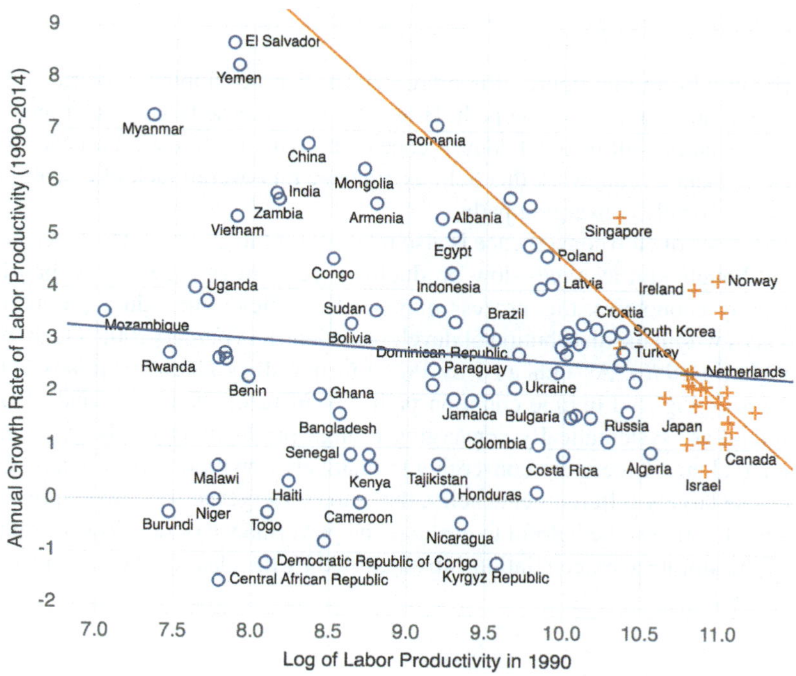

Fig. 1.4 Are there any signs of labor productivity convergence across countries?

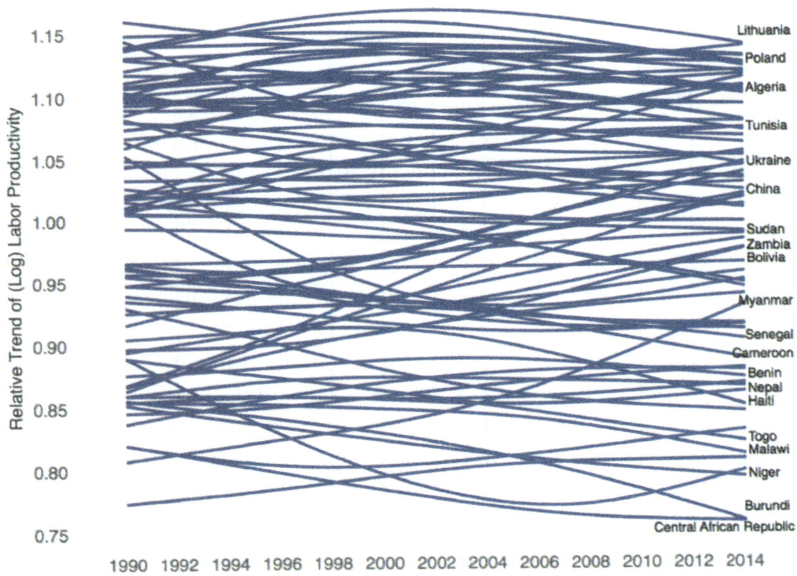

Fig. 1.5 Lack of convergence in developing countries

of each year. From this figure, it is more evident that developing countries do not converge to an overall common path. However, across these largely heterogeneous paths, there could still be local convergence clubs. That is the main purpose of this book, to present a framework that help us organize this overall lack of convergence into a set of locally converging paths.

An analysis of club convergence is also relevant for the group of developed countries. Although—on average—low productivity economies grow faster than high productivity economies, this process may not be sufficient to reduce productivity differences within the distribution of developed countries (Furceri 2005; Quah 1993; Sala-i-Martin 1996). As indicated by Fig. 1.6, not all productivity trends appear to be converging to a unique common path. For instance, Norway, Saudi Arabia, and Ireland are systematically converging to high productivity levels. In contrast, Japan, Israel, and Greece are converging to relatively low productivity levels. Thus, given these largely different tendencies, the club convergence framework presented in this book will also be helpful to evaluate the performance of developed countries beyond the simple average relationship suggested by the classical beta convergence framework.

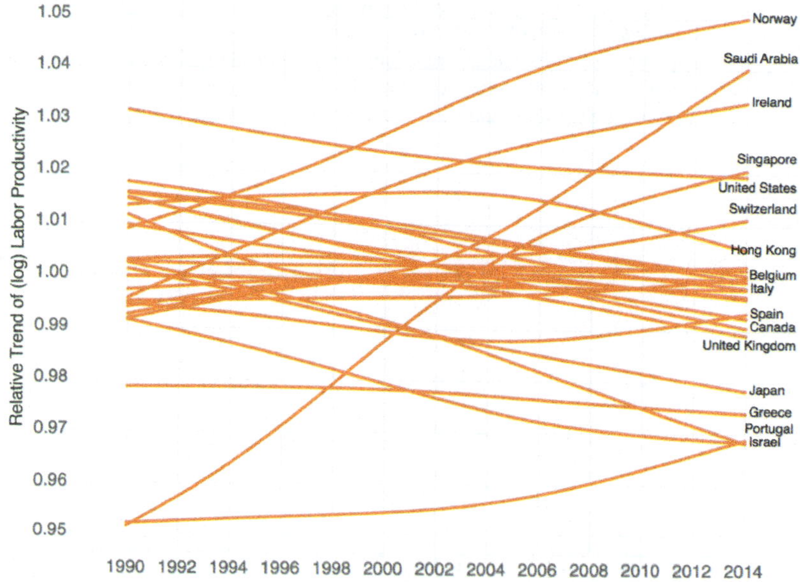

Fig. 1.6 Heterogeneous convergence dynamics in developed countries

References

Barro R, Sala-i-Martin X (1992) Convergence. J Polit Econ 100(2):223–251. https://doi.org/10.1086/261816

Baumol WJ (1986) Productivity growth, convergence, and welfare: what the long-run data show. Am Econ Rev 76(5):1072–1085. https://doi.org/10.2307/1816469

Durlauf, S. N., Johnson, P. A., & Temple, J. R. W. (2009). The Econometrics of Convergence. In T

Furceri D (2005) Beta and sigma-convergence: a mathematical relation of causality. Econ Lett 89(2):212–215

Islam N (2003) What have we learnt from the convergence debate? J Econ Surv 17(3):309–362. https://doi.org/10.1111/1467-6419.00197

Johnson P, Papageorgiou C (2020) What remains of cross-country convergence? J Econ Liter 58(1):129–175. https://doi.org/10.1257/jel.20181207

Patterson MK (eds) Palgrave handbook of econometrics. Palgrave Macmillan, London, pp 1087–1118

Phillips P, Sul D (2007a) Some empirics on economic growth under heterogeneous technology. J Macroecon 29(3), 455–469. https://doi.org/10.1016/j.jmacro.2007.03.002

Phillips P, Sul D (2007b) Transition modeling and econometric. Econometrica 75(6):1771–1855

Phillips P, Sul D (2009) Economic transition and growth. J Appl Econ 24(7):1153–1185. https://doi.org/10.1002/jae

Pittau M, Zelli R, Johnson P (2010) Mixture models, convergence clubs, and polarization. Rev Income Wealth 56(1):102–122

Quah D (1993) Galton's fallacy and tests of the convergence hypothesis. Scandinavian J Econ 95(4):427–443. https://doi.org/10.2307/3440905

Quah D (1997) Empirics for Growth and Distribution: Stratification, Polarization, and Convergence Clubs. J Econ Growth 2(1):27–59. https://doi.org/10.1023/A:1009781613339

Sala-i-Martin X (1996) The classical approach to convergence analysis. Econ J 106(July):1019–1036

Chapter 2
Measuring Labor Productivity and Its Proximate Sources

Abstract This chapter describes a methodological approach to measure labor productivity, capital accumulation, and aggregate efficiency across countries. This approach is based on the well-known production function framework of the economic growth literature. The chapter also describes the construction of the database that is used in the rest of the book. Countries are divided into two groups (developed and developing) based on their income status in 1990. This classification results in 26 developed countries and 82 developing countries, which are evaluated over the 1990–2014 period. Finally, as an overview of the data, the chapter also illustrates the dynamics of each variable based on the quantiles of the cross-country distribution and their evolution.

Keywords Labor productivity · Physical capital · Human capital · Developed countries · Developing countries

2.1 A Production Function Approach

Hall and Jones (1999) study the proximate determinants of labor productivity (output per worker) across countries using a simple production function approach. They start by laying out the components of a Cobb-Douglas production function:

$$Y_{it} = K_{it}^{\alpha} \left(A_{it} H_{it}\right)^{1-\alpha}, \tag{2.1}$$

where Y_{it} is the total output (production) of country i at time t, K_{it} is the stock of physical capital, H_{it} is the stock of human capital, A_{it} is a measure of aggregate efficiency, and α is a parameter that represents the elasticity of output with respect to physical capital. The variable A_{it}, which represents the efficiency with which inputs are used, is known in the literature under different names: total factor productivity, multi-factor productivity, Solow residual, ideas, and the "measurement of our ignorance". In this book, let us refer to A_{it} as aggregate efficiency because it embodies a more general and intuitive notion of the production process. That is, the production

C. Mendez, *Convergence Clubs in Labor Productivity and its Proximate Sources*, SpringerBriefs in Economics, https://doi.org/10.1007/978-981-15-8629-3_2

of total output (Y_{it}) depends on the levels of inputs (K_{it} and H_{it}) and the efficiency with which those inputs are used (A_{it}).

By dividing each side of Eq. (2.1) by the number of workers, L_{it}, we can derive the following production relationship:

$$\frac{Y_{it}}{L_{it}} = \left(\frac{K_{it}}{Y_{it}}\right)^{\alpha/(1-\alpha)} \frac{H_{it}}{L_{it}} A_{it}, \tag{2.2}$$

where Y_{it}/L_{it} is output per worker (labor productivity), K_{it}/Y_{it} is the ratio of physical capital to output, H_{it}/L_{it} is the amount of human capital per worker, and A_{it} still represents the aggregate level of efficiency in the economy. As suggested by Hall and Jones (1999), Hsieh and Klenow (2010), Klenow and RodríÂguez-Clare (1997), Mankiw et al. (1992), using the ratio of physical capital to output as opposed to the ratio of physical capital to labor is more convenient and has a more natural interpretation for long-run analysis. Specifically, the ratio of physical capital to output is proportional to the rate of investment along a balanced growth path. It also helps us to control for the indirect effects on physical capital accumulation that may occur when there is an exogenous change in aggregate efficiency.[1]

2.2 A Database to Study Labor Productivity Across Countries

Largely based on the Penn World Table 9.0, Feenstra et al. (2015), Daude and Fernandez-Arias (2010) and Daude and Fernandez-Arias (2017) have constructed and updated a database that includes series for labor productivity (GPD per worker), physical capital, human capital, and aggregate efficiency. For some countries, the database covers the 1960–2014 period. However, the time series for many developing countries (particularly African countries) are only available since the 1970 and 1980s. Thus, to maximize the number of developing countries in the sample with balanced series for all variables, the 1990–2014 period is selected. Over this time horizon, 108 countries have complete time series for the main variables of the analysis.

Based on the original time series of Daude and Fernandez-Arias (2017), the following variables have been added:

- An indicator variable that classifies countries into developed and developing. A country is classified as developed if it is included in the high-income country list of the World Bank in 1990. All remaining countries are classified as developing countries.[2] Given this classification criteria, there are 26 developed countries and

[1] See Caselli (2005) for a comprehensive production analysis based on the ratio of physical capital to labor.

[2] Although Taiwan is not included in the list of the World Bank, it has been classified as a developed country based on its high income level of 1990.

82 developing countries in the final sample. See Appendix A and B for the list of developed and developing countries included in the sample.

- Long run trends of labor productivity, (physical) capital-output ratio, human capital per worker, and aggregate efficiency. These variables have been estimated using the method of Hodrick and Prescott (1997) with a smoothing parameter of 400. Although other values for the smoothing parameter have been suggested in the literature (Uhlig and Ravn 2002) and there is a current debate (Hamilton 2018) on the usefulness of this method, Phillips and Sul (2009) have used this smoothing parameter in their seminal study on convergence clubs.

2.2.1 Measuring Labor Productivity

Labor productivity is measured as output per worker. Output is measured as total real domestic product (GDP) and, to ensure comparability, it is measured taking into account purchasing power differences across countries. The original data source for output as well as the number of workers is the Penn World Table 9.0. Ideally, one would like to measure labor productivity as output per hour. Data on the number of hours worked, however, are unavailable for many developing countries.

Figures 2.1 and 2.2 provide a first overview of the cross sectional (quantile) dynamics and potential convergence patterns for (log) labor productivity in developed and developing countries, respectively. In both figures, it is difficult to observe signs of productivity convergence. Only in developed countries (Fig. 2.1), a reduction in productivity disparities is observed during the 1996–2000 sub-period. Since then, however, an increasing gap between the top and the bottom of the distribution has become more evident. In developing countries (Fig. 2.2), the quantiles of the productivity distribution look almost parallel. The bottom of the distribution is far away from the top and there is not any sign of convergence among quantiles.

2.2.2 Measuring the (Physical) Capital-Output Ratio

Physical capital stock series are also taken from the Penn World Table (hereafter referred as PWT) 9.0. Series of total investment are needed to compute the series of physical capital. Following Easterly and Levine (2001), the perpetual inventory method is used to construct the capital stock as follows:

$$K_t = K_{t-1}(1 - \delta) + I_t$$

where K_t is the stock of capital in the year t, I is investment and δ is the depreciation rate. In the PWT 9.0, depreciation rates vary among different types of assets.[3] To

[3] See Feenstra et al. (2015) for further details.

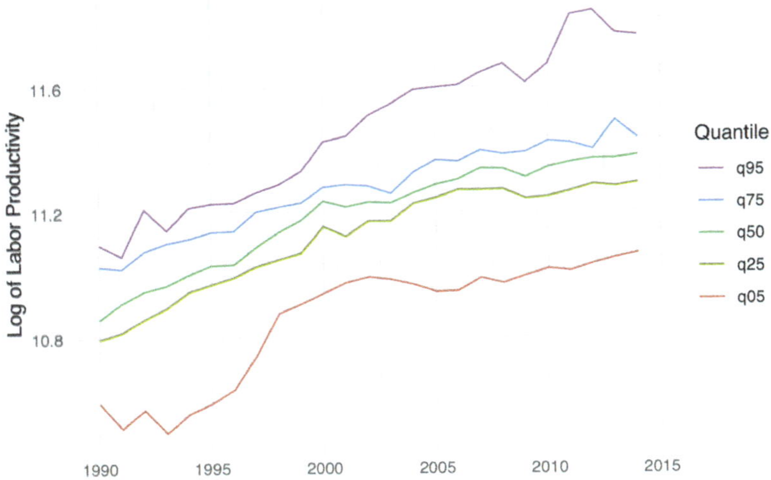

Fig. 2.1 Quantile dynamics of (log) labor productivity in developed countries

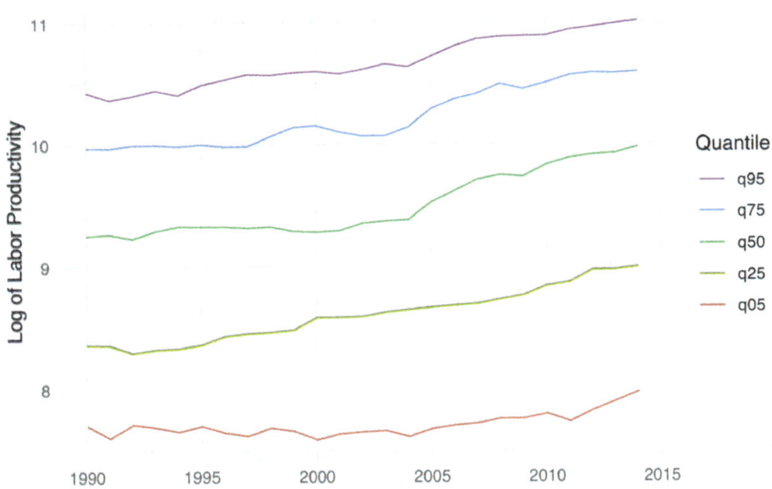

Fig. 2.2 Quantile dynamics of (log) labor productivity in developing countries.eps

obtain the ratio of physical capital to output, K_t is divided by real GDP. To simplify the presentation of results, let us refer to the ratio of physical capital to output, K_{it}/Y_{it}, simply as the capital-output ratio.

Figures 2.3 and 2.4 provide a first overview of the cross sectional (quantile) dynamics and potential convergence patterns for (log) capital-output ratio in developed and developing countries, respectively. In developed countries, the capital-output ratio shows larger cyclical fluctuations. Despite these fluctuations, there are no signs of

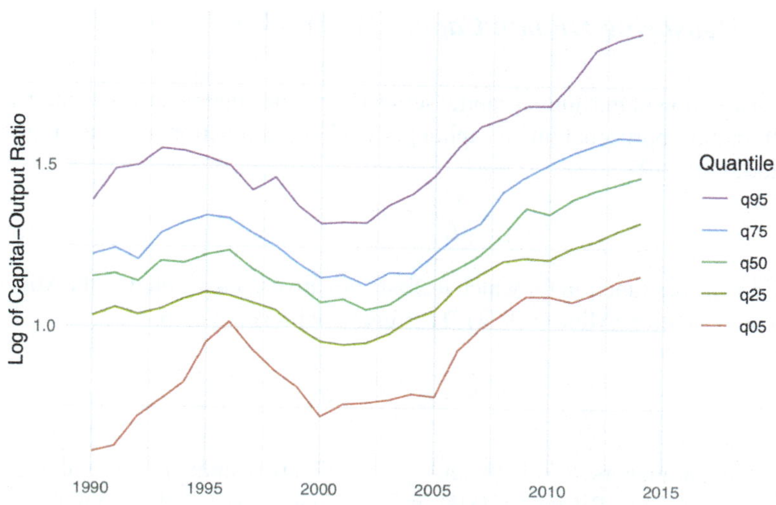

Fig. 2.3 Quantile dynamics of (log) capital-output ratio in developed countries

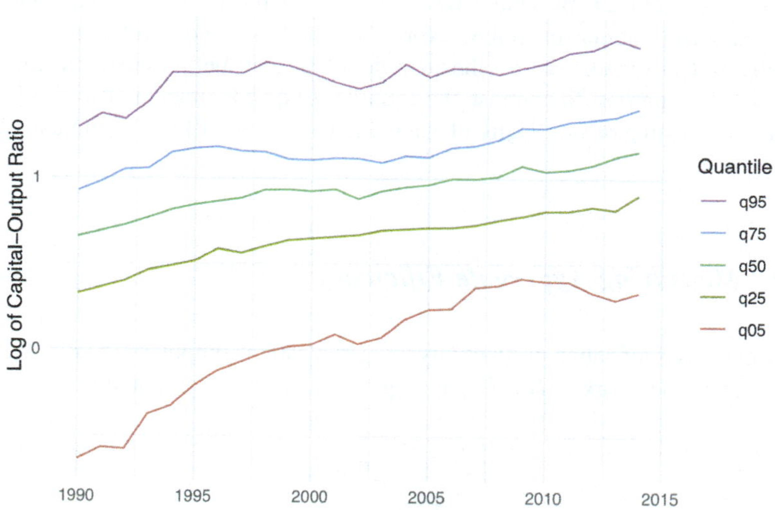

Fig. 2.4 Quantile dynamics of (log) capital-output ratio in developing countries

convergence across quantiles over the entire 1990–2014 period. In contrast, developing countries show some signs of convergence in their capital-output ratios. The bottom of the distribution, in particular the fifth quantile, has been catching up with the rest of the distribution. Thus, cross-country disparities in the capital-output ratio are lower in the year 2014.

2.2.3 Measuring Human Capital Per Worker

The construction of the human capital series follow the approach of Bils and Klenow (2000). In this approach, human capital per worker is a function of the average years of schooling:

$$\frac{H_{it}}{L_{it}} = e^{\phi(S_{it})},$$

where $\phi(S_{it})$ is a function for which its first derivative is interpreted as the Mincerian return on education (Mincer 1974). The functional form of $\phi(S_{it})$ is

$$\phi(S_{it}) = \frac{\theta}{1-\psi} S_{it}^{1-\psi},$$

where the parameters $\theta = 0.18$ and $\psi = 0.37$ are estimated from the data of Psacharopoulos and Patrinos (2004) and the data on years of schooling, S_{it}, are from Barro and Lee (2013).[4]

Figures 2.5 and 2.6 provide a first overview of the cross sectional (quantile) dynamics of (log) human capital per worker in developed and developing countries, respectively. Convergence across the quantiles of the distribution seems more evident in developed countries. In contrast, the quantiles of developing countries are largely parallel, suggesting strong signs of stagnation and thus a lack of human capital convergence.

2.2.4 Measuring Aggregate Efficiency

Given the series of labor productivity, Y_{it}/L_{it}, capital output ratio, K_{it}/Y_{it}, and human capital per worker, H_{it}/L_{it}, aggregate efficiency is computed as:

$$A_{it} = \frac{\frac{Y_{it}}{L_{it}}}{\left(\frac{K_{it}}{Y_{it}}\right)^{\alpha/(1-\alpha)} \frac{H_{it}}{L_{it}}},$$

where the elasticity parameter α is set to $1/3$ following the cross-country findings of Gollin (2002). Although this measurement approach of aggregate efficiency is conceptually similar to the growth accounting framework proposed by Solow (1957), there is an important difference: aggregate efficiency is measured in levels as opposed to growth rates.

[4] Alternatively, the series of human capital per worker can be calculated using the methodological approach of Hall and Jones (1999) with data from Psacharopoulos (1994). Daude and Fernandez-Arias (2017), however, indicate that estimations based on this approach are outdated, and thus do not represent current schooling progress in the world.

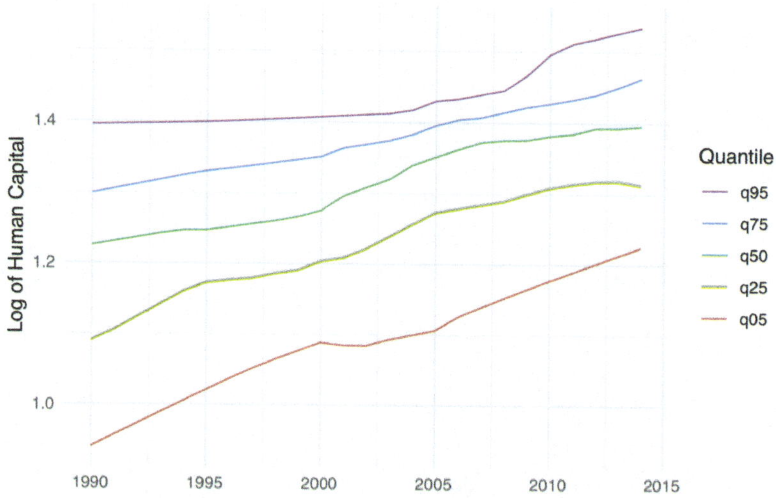

Fig. 2.5 Quantile dynamics of (log) human capital per worker in developed countries

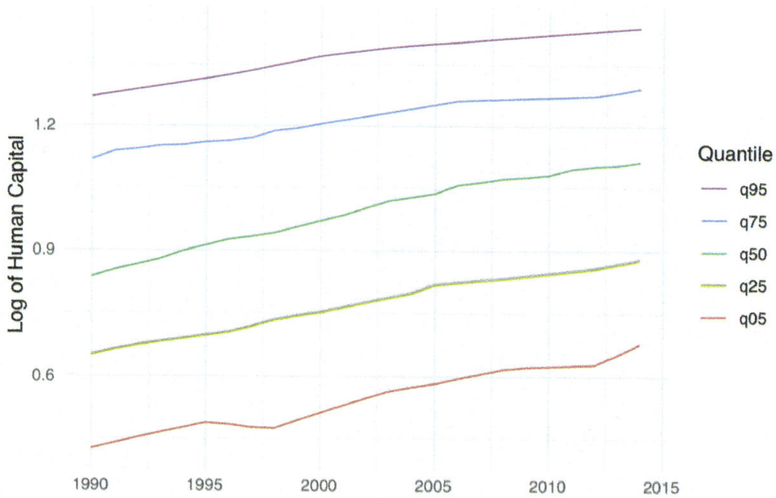

Fig. 2.6 Quantile dynamics of (log) human capital per worker in developing countries

Figures 2.7 and 2.8 provide a first overview of the cross sectional (quantile) dynamics of (log) aggregate efficiency in developed and developing countries, respectively. Similar to human capital, convergence across quantiles seems more noticeable in developed nations. Particularly over the 1990–2000 period, countries at the bottom of the distribution were catching up. Nevertheless, since 2005, the quantiles are mostly parallel, suggesting a lack of efficiency convergence in more recent years. In developing countries, the dynamics of the quantiles are largely parallel over the

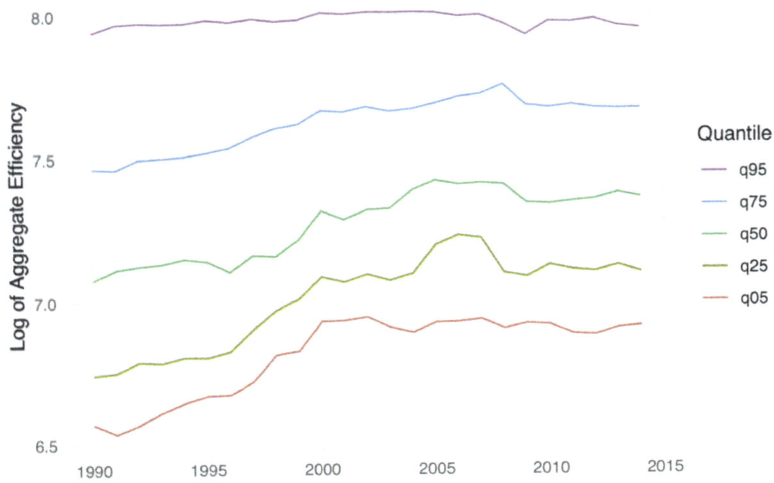

Fig. 2.7 Quantile dynamics of (log) aggregate efficiency in developed countries

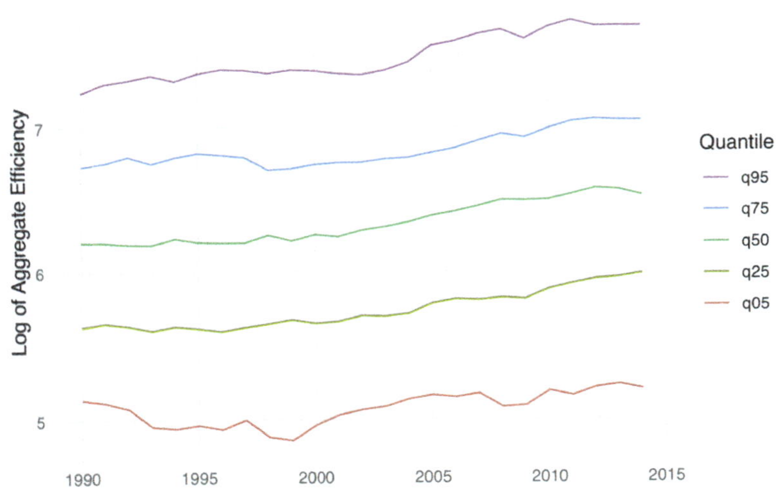

Fig. 2.8 Quantile dynamics of (log) aggregate efficiency in developing countries

entire 1990-2014 period. In addition, efficiency differences across countries have slightly increased over time.

The overall of message of the figures of this chapter is that the dynamics of labor productivity, capital accumulation, and aggregate efficiency are highly heterogeneous across countries and over time. Motivated by these findings, the next chapter presents a framework that will help us organize the individual transition paths of countries. The ultimate purpose of this framework would be to identify clubs of countries in which transition paths are converging to a common long-run equilibrium.

References

Barro RJ, Lee JW (2013) A new data set of educational attainment in the world, 1950–2010. J Dev Econ 104:184–198

Bils M, Klenow PJ (2000) Does schooling cause growth? Am Econ Rev 90(5):1160–1183

Caselli F (2005) Accounting for cross-country income differences. In: Philippe A, Durlauf S (eds) Handbook of economic growth, 1st edn, Vol 1. Elsevier, Amsterdam, pp 679–741. https://doi.org/10.1016/S1574-0684(05)01009-9

Daude C, Fernandez-Arias E (2010) On the role of productivity and factor accumulation in economic development in latin america and the caribbean. Int Am Dev Bank. https://doi.org/10.2139/ssrn.1817273

Daude C, Fernandez-Arias E (2017) On the role of productivity and factor accumulation in economic development in latin america and the caribbean. Int Am Dev Bank. https://doi.org/10.2139/ssrn.1817273

Easterly W, Levine R (2001) What have we learned from a decade of empirical research on growth? It's not factor accumulation: stylized facts and growth models. World Bank Econ Rev 15(2):177–219

Feenstra RC, Inklaar R, Timmer MP (2015) The next generation of the penn world table. Am Econ Rev 105(10):3150–3182. https://doi.org/10.1257/aer.20130954

Gollin D (2002) Getting income shares right. J Polit Econ 110(2):458–474

Hall R, Jones C (1999) Why do some countries produce so much more output per worker than others? Q J Econ 114(1):83–116. https://doi.org/10.1162/003355399555954

Hamilton JD (2018) Why you should never use the Hodrick-Prescott filter. Rev Econ Stat 100(5):831–843

Hodrick RJ, Prescott EC (1997) Postwar US business cycles: an empirical investigation. J Money Credit Banking 1–16

Hsieh CT, Klenow P (2010) Development accounting. Am Econ J Macroecon 2(1):207–223. https://doi.org/10.1257/mac.2.1.207

Klenow P, Rodríguez-Clare A (1997) The neoclassical revival in growth economics: has it gone too far? NBER Macroecon Ann 12:62–103. https://doi.org/10.2307/3585220

Mankiw G, Romer D, Weil D (1992) A contribution to the empirics of 26 Chapter 8. List of developed countries and their convergence clubs growth. Q J Econ 107(2):407–437

Mincer J (1974) Schooling, experience, and earnings. NBER Books

Phillips P, Sul D (2009) Economic transition and growth. J Appl Econ 24(7):1153–1185. https://doi.org/10.1002/jae

Psacharopoulos G (1994) Returns to investment in education: a global update. World Dev 22(9):1325–1343

Psacharopoulos G, Patrinos HA (2004) Returns to investment in education: a further update. Edu Econ 12(2):111–134

Solow R (1957) Technical change and the aggregate production function. Rev Econ Stat 312–320

Uhlig H, Ravn M (2002) On adjusting the Hodrick-Prescott filter for the frequency of observations. Rev Econ Stat 84(2):371–376

Chapter 3
A Modern Framework to Study Convergence

Abstract This chapter presents a modern methodological framework to study economic convergence across countries. It starts with a brief overview of the classical convergence framework and its limitations. Next, it presents a modern club convergence framework that encompasses key economic phenomena such as technological heterogeneity and multiple equilibria. Econometrically, this framework is based on a non-linear dynamic factor model that evaluates the hypothesis that all countries would eventually converge to a common steady-state growth path. If this hypothesis is rejected, then a clustering algorithm for panel data is applied to identify local convergence clubs. The chapter concludes with a selected overview of the club convergence literature and a tutorial on how to implement the club convergence framework in the statistical software Stata.

Keywords Classical convergence · Club convergence · Technological heterogeneity · Multiple equilibria · Stata

3.1 Classical Beta Convergence Approach: Technological Homogeneity and a Unique Equilibrium

Based on different statistical frameworks and assumptions, there are different concepts of economic convergence. Among them, beta convergence is the most popular. In the early 1990s, Barro and Sala-i-Martin (1992) proposed this concept in an attempt to test the predictions of the neoclassical growth model. Empirically, this concept of convergence is commonly summarized by the inverse relationship between the initial level of a variable and its subsequent growth rate. If such relation holds, and taking as an example the income differences across countries, it would imply that initially poor countries grow faster than initially rich countries. Thus, there is a process of convergence across countries. Friedman (1992) and Quah (1993) criticized this approach as an statistical illusion and suggested that the true notion of convergence should be related to the variance and dynamics of the entire income distribution.

C. Mendez, *Convergence Clubs in Labor Productivity and its Proximate Sources*, SpringerBriefs in Economics, https://doi.org/10.1007/978-981-15-8629-3_3

One of the main purposes of the beta convergence approach has been testing the predictions of the neoclassical growth model. In this model, countries are mainly characterized by diminishing marginal returns to capital accumulation and by a homogeneous rate of technological progress. These two conditions allow countries to convergence to a unique equilibrium in the long run.[1]

In the standard neoclassical growth model, transitional dynamics of log output per worker are expressed as

$$\log y_{it} = \log \tilde{y}_i^* + \log A_0 + \left[\log \tilde{y}_{i0} - \log \tilde{y}_i^*\right] e^{-\beta t} + xt$$

where $\tilde{y}_{i0}$ and $\tilde{y}_i^*$ are the initial and steady-state levels of effective output per worker, A_0 is initial level of technology, x is the growth rate of technology, and β is the speed of convergence. Note that the speed of convergence, β, and the growth rate of technology, x, are common (homogeneous) across countries, thus convergence across countries is an expected result in this economic model.

3.2 Modern Club Convergence Approach: Technological Heterogeneity and Multiple Equilibrium

Phillips and Sul (2007a) derive a growth model with heterogeneous technological progress. Transitional dynamics of log output per worker are expressed as

$$\log y_{it} = \log \tilde{y}_i^* + \left[\log \tilde{y}_{i0} - \log \tilde{y}_i^*\right] \mathrm{e}^{-\beta_{it} t} + x_{it} t$$

where

$$\beta_{it} = \beta_i - \frac{1}{t} \log \left\{ 1 - d_{i1} \int_0^t \mathrm{e}^{\beta_i m} \left(x_{im} - x\right) \mathrm{d}m \right\},$$

$$d_{i1} = \left(\log k_{i0} - \log k_i^*\right)^{-1},$$

k_{i0} and k_i^* are the initial and steady-state levels of effective capital per worker. Note that in this model the speed of convergence, β_{it}, and the growth rate of technology, x_{it}, are now heterogeneous across countries as well as over time. Thus, in this more general model, convergence towards a unique equilibrium is not guaranteed.

To empirically evaluate this growth model with heterogeneous technology, Phillips and Sul (2007b) proposed the following nonlinear dynamic factor model

$$\log y_{it} = g_{it} + x_{it} t, \tag{3.1}$$

[1]In addition to common rate of technological progress, convergence is conditional on countries sharing similar population growth rates, savings rates, and depreciation rates. Also, the production function is consistent with the conditions of Inada (1963).

where g_{it} represents the transitional dynamics for capital per worker and x_{it} represents the idiosyncratic time paths of technological change. Alternatively, Eq. (3.1) can be restated as:

$$\log y_{it} = \left(\frac{g_{it} + x_{it}}{\mu_t} \right) \mu_t = \delta_{it} \mu_t, \tag{3.2}$$

where δ_{it} represents an idiosyncratic component and μ_t represents a common component. From an economic standpoint, δ_{it} describes the transition path of each economy, and μ_t describes the equilibrium growth path that is common to all economies. From an statistical standpoint, Eq. (3.2) is a dynamic factor model where the idiosyncratic component, δ_{it}, is a factor-loading coefficient that represents the distance between individual behavior ($\log y_{it}$) and common behavior (μ_t).

Next, Phillips and Sul (2007b) suggest the following semi-parametric econometric specification to describe the dynamics of the idiosyncratic component (δ_{it}):

$$\delta_{it} = \delta_i + \frac{\sigma_i \xi_{it}}{log\,(t)\, t^{\alpha}},$$

where δ_i only varies across economies and ξ_{it} is a weakly time dependent process with mean 0 and variance 1 across economies. Given this setting, convergence is achieved when all economies move to the same transition path. That is,

$$\lim_{t \to \infty} \delta_{it} = \delta \text{ and } \alpha \geq 0. \tag{3.3}$$

To evaluate this hypothesis Phillips and Sul (2007b) define a relative transition parameter, h_{it}, as:

$$h_{it} = \frac{\log y_{it}}{\frac{1}{N}\sum_{h=1}^{N} \log y_{it}} = \frac{\delta_{it}}{\frac{1}{N}\sum_{i=1}^{N} \delta_{it}}.$$

This parameter helps remove the common component, μ_t, from Eq. (3.2) by dividing the observed variable by its cross-sectional mean in each year. In the long run ($t \to \infty$), the convergence hypothesis defined in Eq. (3.3) is equivalent to

$$H_t = \frac{1}{N}\sum_{i=1}^{N} (h_{it} - 1)^2 \to 0.$$

This definition is particularly intuitive as it indicates that convergence occurs when the cross-sectional variance tends to zero, $H_t \to 0$. Or, alternatively, when the relative transition parameter tends to unity, $h_{it} \to 1$.

Finally, to empirically test this hypothesis of convergence, Phillips and Sul (2007b) propose the following log-t regression model:

$$log\left(\frac{H_1}{H_t}\right) - 2log\,\{log\,(t)\} = a + b\,log\,(t) + \epsilon_t$$
$$\text{for } t = [rT], [rT] + 1, \ldots, T \text{ with } r > 0, \tag{3.4}$$

where $[rT]$ is the initial observation in the regression, which implies that the first fraction of the data (that is, r) is discarded. Based on Monte Carlo simulations, Phillips and Sul (2007b) suggest to set $r = 0.3$ when the time horizon is less or equal to 50 years.

Figure 3.1 provides an illustration that facilitates the interpretation of Eq. (3.4). The sign of coefficient b indicates whether the ratio of cross-sectional variance $\frac{H_1}{H_t}$ is increasing or decreasing over time. When b is positive (negative), the cross-sectional variance, H_t, tends to be smaller (larger) than the initial cross-sectional variance, H_1. Thus, when this coefficient is positive (negative) and statistically significant, the log-t regression model would suggest convergence (divergence) among the cross-sectional units of the sample.

In addition to its sign, the magnitude of b provides further information to classify the convergence process. When $0 \leq b < 2$, the model suggests convergence in growth rates (that is, relative convergence). When $b \geq 2$, the model suggests convergence in levels (that is, absolute convergence). Finally, an indicator of the speed of convergence can be computed as $b/2$.

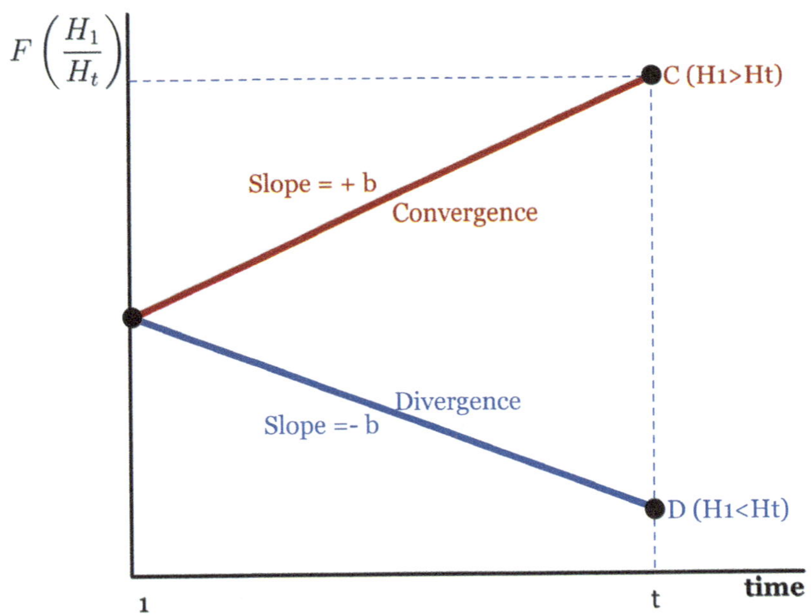

Fig. 3.1 A visual summary of the convergence test

To evaluate the statistical significance of b, Phillips and Sul (2007b) propose conventional inferential procedure. It is based on a one-sided t test with a limit distribution:

$$t_b = \frac{\widehat{b} - b}{s_b} \Rightarrow N(0, 1)$$

where

$$s_b^2 = \widehat{\text{lvar}}\left(\widehat{\varepsilon}_t\right)\left\{\sum_{t=[rT]}^{T}\left(\log(t) - \frac{1}{T - [rT] + 1}\sum_{t=[rT]}^{T}\log(t)\right)^2\right\}^{-1},$$

and $\widehat{\text{lvar}}\left(\widehat{\varepsilon}_t\right)$ is a heteroskedastic and autocorrelated estimate constructed from the regression residuals. In this inferential setting, the null hypothesis of convergence is rejected when $t_b < -1.65$.

3.3 Finding Convergence Clubs: A Clustering Algorithm For Panel Data

When the null hypothesis of global convergence is rejected by the data, multiple local convergence clubs are likely to exist. Figure 3.2 illustrates this phenomenon for four hypothetical countries. In this figure, none of the countries are converging to the global (cross-country) average. Instead, there are two local equilibria, one above and the other below the average.

To identify local convergence clubs, Phillips and Sul (2007b) proposed a data-driving clustering algorithm. Based on the log-t convergence test of Eq. (3.4), this clustering algorithm evaluates the transition paths of all economies as well as the transition paths of the emerging clubs. The five steps of this algorithm are briefly summarized as follows.

Step 1: Cross-sectional ordering
Economies are sorted in decreasing order according to the observed values in the last period. In cases where substantial time-series volatility is observed, the sorting can be done according to averaged values across the last half or the last third of the time horizon.

Step 2: Core group formation
A group of k highest economies is selected for $N > k \geq 2$. The criteria for selecting the optimal size, k^*, of this group is

$$k^* = \arg\max_k \{t_k\} \quad \text{subject to } \min\{t_k\} > -1.65$$

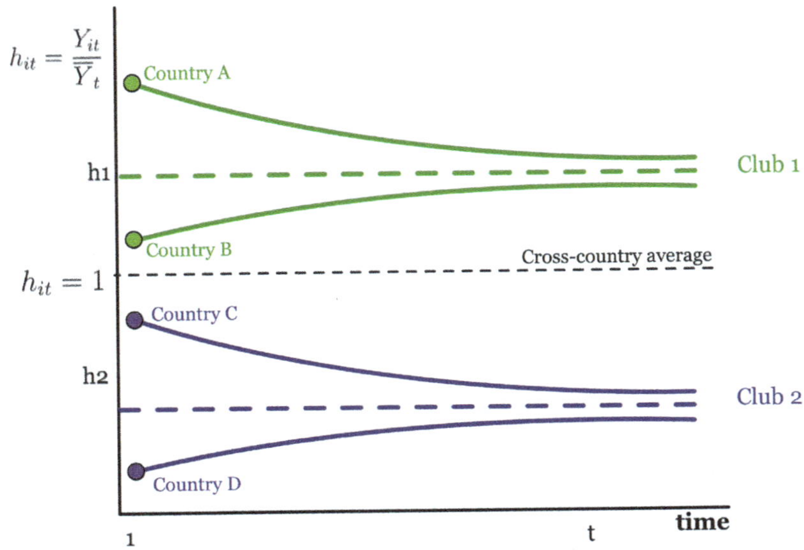

Fig. 3.2 A visual summary of the convergence clubs

where t_k refers to the one-side t-statistic that is needed to evaluate statistical significance of the convergence test of Eq. (3.4). If $t_k > -1.65$ does not hold for $k = 2$, then the economy with the highest value can be dropped and the algorithm can be reevaluated again for the rest of sample.

Step 3: Sieve economies for club membership
Economies not belonging to the core group form a complementary group G^c. Add one economy from G^c at each time to the core group k^* and run the log-t test (Eq. (3.4)). A new group is formed when the t-statistic is greater than zero.

Step 4: Recursion and stopping rule
The log-t test (Eq. (3.4)) is applied for the remaining economies. If convergence is rejected, Steps 1 to 3 are repeated. If no core group is found, then the remaining economies are labeled as divergent economies and the algorithm stops.

Step 5: Club merging
The the log-t test is performed for all pairs of initial clubs. The merging procedure is iterative. That is, the log-t test for the initial clubs 1 and 2 is run; and if they fulfill the convergence test jointly, they should be merged into a new club 1. This merging procedure continues for the other two clubs until the convergence test is rejected.

3.4 Brief Overview of the Club Convergence Literature

A comprehensive survey of the club convergence literate is beyond the scope of this brief monograph.[2] Nevertheless, this section provides a succinct overview of both seminal papers and more recent contributions that have applied the approach of Phillips and Sul (2007b) to study convergence across countries. Phillips and Sul (2007a) develop a relative (club) convergence framework as an augmented version of the Solow growth model. They emphasize the importance of technological heterogeneity in economic growth and provide some empirical illustrations of the dynamics of the transitional parameter, h_{it}, for GDP per capita. Phillips and Sul (2009) provide further empirical applications for studying income convergence in the context of regional US data, OECD data, and Penn World Table data. Results indicate convergence for both the US regional data and the OECD data. In contrast, data from the Penn World Table for a sample of developed and developing countries indicate four convergence clubs and a group of diverging economies.

After the seminal papers of Phillips and Sul 2007a, b and 2009, the club convergence framework has been used in different contexts beyond income per capita. For instance, Panopoulou and Pantelidis (2009) study convergence in carbon dioxide emissions among 128 countries over the 1960–2003 period. They find two separate convergence clubs with some countries transitioning between clubs. Antzoulatos et al. (2011) study financial system convergence across a large set of industrial and developing countries. Overall convergence is reject across 13 indicators of financial development. More recently, Basel et al. (2020) study club convergence in an multidimensional index of development across 102 countries over the 1996-2015 period.[3] Results indicate the existence of four clubs with some evidence of transition during the financial crisis of 2008.

Other studies have evaluated club convergence hypothesis within continents or groups of countries. For instance, Apergis et al. (2010) study club convergence in GDP per capita and its proximate sources across countries of the European Union over the 1980–2004 period. Their findings suggest two distinct convergence clubs with considerable heterogeneity in the proximate determinants of GDP per capita.[4] Barrios et al. (2019) study club convergence in GDP per capita for a sample of 17 Latin American countries over the 1990–2014 period. Their findings suggest four convergence clubs with largely parallel trends. Tam (2018) studies club convergence in GDP per capita across 19 Asian countries over the 1970–2014 period. Results indicate overall growth convergence, but not level convergence. Ulucak et al. (2020) studies club convergence in ecological footprint and its sub-components for a sample

[2] See Johnson and Papageorgiou (2020) for a recent survey of the cross-country convergence literature.

[3] The index consists of seven indicators: education, health, access to water and sanitation, energy use, environment, standard of living, and good governance.

[4] The proximate determinants of GDP per capita are those of a standard growth accounting framework.

of 23 countries from Sub-Saharan Africa over the 1961–2014 period. Results indicate a lack of overall convergence in economical footprint and most of its sub-components.

3.5 Convergence Test and Identification of Clubs Using Stata

Du (2017) introduced a Stata package to perform the econometric convergence analysis and club clustering algorithm of Phillips and Sul (2007b). Although the package is well documented and easy to use, it does not include commands to create figures or export result tables. In what follows, the basic use of the package is described with some additional pieces of code to automate the creation of figures and tables.[5]

```
*-------------------------------------------------------
***************** Code and data*************************
*-------------------------------------------------------

*   Available at: https://bit.ly/mendez2020code

*-------------------------------------------------------
```

The code below installs the convergence clubs package and its dependencies. It is important to note that Stata 12.1 or higher is needed to run the convergence clubs package. In addition, to export the results to excel, Stata 14.2 or higher is needed to use the `putexcel` command. Finally, note that this installation needs to be done only once.

```
*-------------------------------------------------------
***************** Install packages*********************
*-------------------------------------------------------
* Install the convergence clubs package (findit st0503_1)
net install st0503_1.pkg, from(http://www.stata-journal.com/software/sj19-1)
* Install package dependencies
ssc install moremata
*-------------------------------------------------------
```

After installing the package, we need to define some global (macro) parameters such as the name of the dataset (for example, `hiYes_log_lp`), the main variable to be studied (for example, `log_lp`), the label of that variable (for example, `Labor Productivity`), the type of cross-sectional unit (for example, `country`), and the type of temporal unit (for example,`year`). Users of this code should carefully check these five parameters as the next steps crucially depend on them to work correctly.

[5]Code and data are available at https://bit.ly/mendez2020code or https://github.com/quarcs-lab/mendez2020-convergence-clubs-code-data.

```
*-------------------------------------------------------
clear all
macro drop _all
set more off

*-------------------------------------------------------
***************** Define five global parameters*********
*-------------------------------------------------------
* (1) Indicate name of the dataset (Example: hiYes_log_lp.dta)
global dataSet hiYes_log_lp
* (2) Indicate name of the variable to be studied (Example: log_lp)
global xVar log_lp
* (3) Write label of the variable (Example: Labor Productivity)
global xVarLabel Labor Productivity
* (4) Indicate cross-sectional unit ID (Example: country)
global csUnitName country
* (5) Indicate temporal unit ID (Example: year)
global timeUnit year
*-------------------------------------------------------
```

To have a record of the written commands and results (excluding the display of figures), let us start a log file. The name of this file is automatically captured from the previously defined parameters.

```
*-------------------------------------------------------
***************** Start log file************************
*-------------------------------------------------------
log using "${dataSet}_clubs.txt", text replace
*-------------------------------------------------------
```

Next, from the current working directory, we load the dataset, which is in a .dta format, and set the structure of the data. Again, we do not have to modify anything from this code as long as the global parameters are correctly defined.

```
*-------------------------------------------------------
***************** Load and set panel data  ***********
*-------------------------------------------------------
** Load data
use "${dataSet}.dta"

* Keep necessary variables
keep id ${csUnitName} ${timeUnit} ${xVar}

* Set panel data
xtset id ${timeUnit}
*-------------------------------------------------------
```

The next piece of code is the most important one of the entire package. It runs the log-t convergence test, the clustering and merge algorithms, and lists the final results in a table. If we are using a log file, all code and results are recorded in the `dataSet_clubs.txt` file. In addition, by using the `putexcel` we can export the results in a table form to excel.

```
*-------------------------------------------------------
****************** Apply PS convergence test  ***********
*-------------------------------------------------------
* (1) Run log-t regression
putexcel set "${dataSet}_test.xlsx", sheet(logtTest) replace
  logtreg ${xVar},  kq(0.333)

ereturn list
matrix result0 = e(res)
putexcel A1 = matrix(result0), names nformat("#.##") overwritefmt

* (2) Run clustering algorithm
putexcel set "${dataSet}_test.xlsx", sheet(initialClusters) modify
  psecta ${xVar}, name(${csUnitName}) kq(0.333) gen(club_${xVar})

matrix b=e(bm)
matrix t=e(tm)
matrix result1=(b \ t)
matlist result1, border(rows) rowtitle("log(t)") format(%9.3f) left(4)
putexcel A1 = matrix(result1), names nformat("#.##") overwritefmt

* (3) Run merge algorithm
putexcel set "${dataSet}_test.xlsx", sheet(mergingClusters) modify
  scheckmerge ${xVar},  kq(0.333) club(club_${xVar})

matrix b=e(bm)
matrix t=e(tm)
matrix result2=(b \ t)
matlist result2, border(rows) rowtitle("log(t)") format(%9.3f) left(4)
putexcel A1 = matrix(result2), names nformat("#.##") overwritefmt

* (4) List final clusters
putexcel set "${dataSet}_test.xlsx", sheet(finalClusters) modify
  imergeclub ${xVar}, name(${csUnitName}) kq(0.333) club(club_${xVar})
  gen(finalclub_${xVar})

matrix b=e(bm)
matrix t=e(tm)
matrix result3=(b \ t)
matlist result3, border(rows) rowtitle("log(t)") format(%9.3f) left(4)
putexcel A1 = matrix(result3), names nformat("#.##") overwritefmt
*-------------------------------------------------------
```

To plot the dynamics of the cross-sectional units and their respective convergence clubs, we first need to re-scale the data based on the cross-sectional average of each year. The code below performs that task. The result of this code is an extended panel dataset (in both `.dta` and `.csv` formats) that includes the list of countries, club membership, and the absolute and relative values of the variable under study.

```
*-------------------------------------------------------
****************** Generate relative variables**********
*-------------------------------------------------------
** Generate relative variable (useful for ploting)
save "temporary1.dta",replace
use  "temporary1.dta"

collapse ${xVar}, by(${timeUnit})
```

```
gen  id=999999
append using "temporary1.dta"
sort id ${timeUnit}

gen ${xVar}_av = ${xVar} if id==999999
bysort ${timeUnit} (${xVar}_av): replace ${xVar}_av = ${xVar}_av[1]
gen re_${xVar} = 1*(${xVar}/${xVar}_av)
label var re_${xVar}  "Relative ${xVar}  (Average=1)"
drop ${xVar}_av
sort id ${timeUnit}

drop if id == 999999
rm "temporary1.dta"

* Order variables
order ${csUnitName}, before(${timeUnit})
order id, before(${csUnitName})

* Export data to csv
export delimited using "${dataSet}_clubs.csv", replace
save "${dataSet}_clubs.dta", replace
*-------------------------------------------------------
```

Given the extended dataset, the code below plots multiple figures and export them as `.pdf` and `.gph` formats. There are three types of plots. First, the relative transition paths of all countries are plotted. This plot is useful as it provides a first graphical overview of dataset. Second, relative transition paths are plotted based on the club classification. Not only a plot for each club is created, but there is also a plot that compares all clubs using a common y-axis. Third, a plot based on within-club averages is also created. It is important to note that the colors and design of figures are based on the `plotplainblind` scheme.[6]

```
*-------------------------------------------------------
***************** Plot the clubs *********************
*-------------------------------------------------------
** (1) All transition paths

xtline re_${xVar}, overlay legend(off) scale(1.6)  ytitle("${xVarLabel}",
size(small)) yscale(lstyle(none)) ylabel(, noticks labcolor(gs10))
xscale(lstyle(none)) xlabel(, noticks labcolor(gs10))  xtitle("")
name(allLines, replace)

graph save   "${dataSet}_allLines.gph", replace
graph export "${dataSet}_allLines.pdf", replace

** (2) Paths for each club

summarize finalclub_${xVar}
return list
scalar nunberOfClubs = r(max)
```

[6]See Bischof (2017) for further information about the graphical scheme. This scheme can be installed by typing the following in the Stata console: `net install gr0070.pkg`, from (http://www.stata-journal.com/software/sj17-3). Activate the scheme by typing: `set scheme plotplainblind`.

```
forval i=1/`=nunberOfClubs' {
    xtline re_${xVar} if finalclub_${xVar} == `i', overlay title("Club `i'",
    size(small)) legend(off) scale(1.5) yscale(lstyle(none))
    ytitle("${xVarLabel}", size(small)) ylabel(, noticks labcolor(gs10))
    xtitle("") xscale(lstyle(none)) xlabel(, noticks labcolor(gs10))
    name(club`i', replace)
    local graphs `graphs' club`i'
    }
graph combine `graphs', ycommon
graph save   "${dataSet}_clubsLines.gph", replace
graph export "${dataSet}_clubsLines.pdf", replace

** (3) Within-club averages

collapse (mean) re_${xVar}, by(finalclub_${xVar} ${timeUnit})
xtset finalclub_${xVar} ${timeUnit}
rename finalclub_${xVar} Club
xtline re_${xVar}, overlay scale(1.6) ytitle("${xVarLabel}", size(small))
yscale(lstyle(none)) ylabel(, noticks labcolor(gs10)) xscale(lstyle(none))
xlabel(, noticks labcolor(gs10))  xtitle("") name(clubsAverages, replace)

graph save   "${dataSet}_clubsAverages.gph", replace
graph export "${dataSet}_clubsAverages.pdf", replace

clear
use "${dataSet}_clubs.dta"
*---------------------------------------------------------
```

The code below exports the list of countries and their club membership to a `.csv` file. This list can be used as a handy reference in the appendix section of a publication.

```
*---------------------------------------------------------
***************** Export list of clubs  ****************
*---------------------------------------------------------
summarize ${timeUnit}
scalar finalYear = r(max)
keep if ${timeUnit} == `=finalYear'
keep id ${csUnitName} finalclub_${xVar}
sort finalclub_${xVar} ${csUnitName}
export delimited using "${dataSet}_clubsList.csv", replace
*---------------------------------------------------------
```

Finally, the code below closes the log file.

```
*---------------------------------------------------------
***************** Close log file*************
*---------------------------------------------------------
log close
*---------------------------------------------------------
```

References

Antzoulatos AA, Panopoulou E, Tsoumas C (2011) Do financial systems converge? Rev Int Econ 19(1):122–136

Apergis N, Panopoulou E, Tsoumas C (2010) Old wine in a new bottle: Growth convergence dynamics in the EU. Atlantic Econ J 38(2):169–181. https://doi.org/10.1007/s11293-010-9219-1

Barrios C, Flores E, Martínez MÁ (2019) Convergence clubs in Latin America. Appl Econ Lett 26(1):16–20. https://doi.org/10.1080/13504851.2018.1433288

Barro R, Sala-i-Martin X (1992) Convergence. J Polit Econ 100(2):223–251. https://doi.org/10.1086/261816

Basel S, Gopakumar KU, Prabhakara Rao R (2020) Testing club convergence of economies by using a broad-based development index. Geo J. https://doi.org/10.1007/s10708-020-10198-0

Bischof D (2017) New graphic schemes for stata: Plotplain and plottig. Stata J 17(3):748–759. https://doi.org/10.1177/1536867X1701700313

Du K (2017) Econometric convergence test and club clustering using stata. Stata J 17(4):882–900. https://doi.org/10.1177/1536867X1801700407

Friedman M (1992) Do Old Fallacies Ever Die ? Published by : American Economic Association Cornmunication. Journal of Economic Literature 30(4):2129–2132

Inada K-I (1963) On a two-sector model of economic growth: Comments and a generalization. Rev Econ Stud 30(2):119–127

Johnson P, Papageorgiou C (2020) What Remains of Cross-Country Convergence? J Econ Liter 58(1):129–175. https://doi.org/10.1257/jel.20181207

Panopoulou E, Pantelidis T (2009) Club convergence in carbon dioxide emissions. Environ Resour Econ 44(1):47–70

Phillips P, Sul D (2007a) Some empirics on economic growth under heterogeneous technology. J Macroecon 29(3):455–469. https://doi.org/10.1016/j.jmacro.2007.03.002

Phillips P, Sul D (2007b) Transition modeling and econometric. Econometrica 75(6):1771–1855

Phillips P, Sul D (2009) Economic transition and growth. J Appl Econ 24(7):1153–1185. https://doi.org/10.1002/jae

Quah D (1993) Galton's Fallacy and Tests of the Convergence Hypothesis. Scandinavian J Econ 95(4):427–443. https://doi.org/10.2307/3440905

Tam PS (2018) Economic transition and growth dynamics in asia: Harmony or discord? Comp Econ Stud 60(3):361–387

Ulucak R, Kassouri Y, Ilkay SÇ, Altıntas H, Garang APM (2020) Does convergence contribute to reshaping sustainable development policies? Insights from sub-saharan africa. Ecol Ind 112:106140

Chapter 4
Convergence Clubs in Labor Productivity

Abstract This chapter studies convergence in labor productivity across developed and developing countries, respectively. Through the lens of a non-linear dynamic factor model, the cross-country dynamics of potential GDP per worker are evaluated over the 1990–2014 period. Results reject the hypothesis that all countries would eventually converge to a common steady-state growth path within their respective groups. Developed countries are characterized by three convergence clubs with largely separating trends. Developing countries are characterized by five convergence clubs. Most clubs are largely below the mean and continue diverging downwards.

Keywords Club convergence · Labor productivity · Non-linear dynamic factor model · Developed countries · Developing countries

4.1 Testing for Overall Convergence

As the variable under study is normalized by the cross-sectional mean of each year, the convergence approach of Phillips and Sul (2007) is also referred as relative convergence (Sul 2019). Figures 4.1 and 4.2 show the transition paths of labor productivity for developed and developing countries, respectively. In these figures, labor productivity of each country is computed as potential (log) GDP per worker divided by the cross-sectional mean of each year. These figures are consistent with the descriptive patterns presented in Chap. 2 in the sense that large productivity differences across countries are observable during the entire 1990-2014 period. From a visual inspection, Figs. 4.1 and 4.2 indicate a high degree of productivity heterogeneity as well as a lack of overall convergence within both groups of countries.

Tables 4.1 and 4.2 formally evaluate the notion of relative convergence based on the log-t convergence test of Phillips and Sul (2007). The log-t test is based on a non-linear dynamic factor model that allows the modeling of transitional heterogeneity. As expected, given the patterns of Figs. 4.1 and 4.2, the null hypothesis of productivity convergence is rejected by the data. In both developed and developing countries, the

C. Mendez, *Convergence Clubs in Labor Productivity and its Proximate Sources*, SpringerBriefs in Economics, https://doi.org/10.1007/978-981-15-8629-3_4

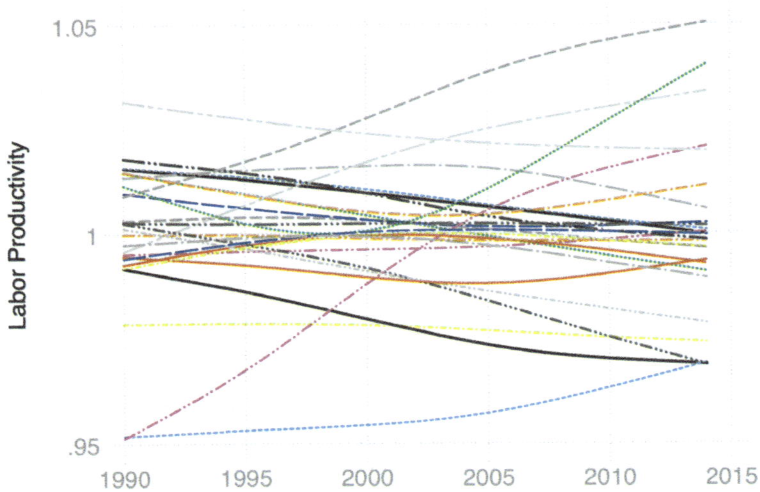

Fig. 4.1 Relative labor productivity in developed countries

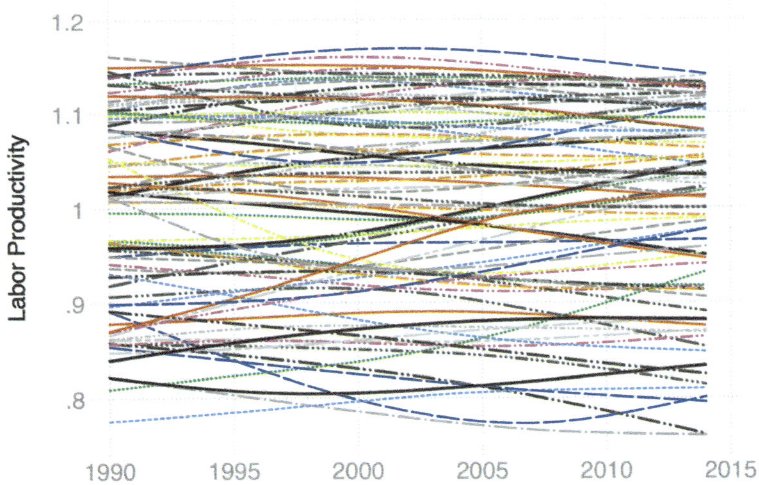

Fig. 4.2 Relative labor productivity in developing countries

coefficient of the log-t regression is negative and statistically significant. Thus, cross-country differences in labor productivity have actually increased over time within both groups of countries.

Table 4.1 Global convergence test: Labor productivity in developed countries

Test	Coeff	SE	T-stat
Log(t)	−1.4	0.049	−28.6

Table 4.2 Global convergence test: Labor productivity in developing countries

Test	Coeff	SE	T-stat
Log(t)	−0.744	0.018	−40.9

4.2 Finding Local Convergence Clubs

Although convergence towards a common equilibrium is rejected by the data, the transition paths of individual countries suggest existence of multiple convergence clubs. For instance, as shown in Fig. 4.1, there may be three groups of countries in 2014. The first group may appear to be located at the top of productivity distribution. The second group may appear to be located around the cross-country mean. And the third group may be located at the bottom of the distribution.

Table 4.3 formally evaluates this intuition by applying the clustering algorithm of Phillips and Sul (2007). Initial results indicate that in developed countries, labor productivity is characterized by four convergence clubs. For each club, the coefficient of the log-t regression is positive and statistically significant. Thus, within each club, labor productivity differences have been decreasing over time.

The magnitude of the coefficient of the log-t regression is also informative for understanding the strength of the convergence process within each club (Johnson 2020). As this coefficient is less than two, Club 1, Club 2, and Club 3 appear to be converging in growth rates. Club 4, however, shows a much larger coefficient, which suggests that the countries belonging to this club are converging in levels.

As noted by Phillips and Sul (2009) and Lyncker and Thoennessen (2017), the convergence clubs algorithm of Phillips and Sul (2007) may over-predict the number of clubs. To handle this issue, it is highly recommended to perform a sequential merge test between the clubs. Table 4.4 shows the results of this test. Although all coefficients are negative, the main criteria for evaluating club convergence is the t-statistic. When this statistic is less than -1.65, the null hypothesis of convergence can be rejected. Following this criteria, the merge between Club 1 and Club 2 and between Club 3 and Club 4 can be rejected by the data. The merge between Club 2 and

Table 4.3 Local convergence test: Labor productivity in developed countries

Statistic	Club1	Club2	Club3	Club4
Coeff	0.206	0.17	0.111	4.59
T-stat	2.287	2.25	2.349	2.46

Table 4.4 Club merge test: Labor productivity in developed countries

Statistic	Club1+2	Club2+3	Club3+4
Coeff	−1.08	−0.058	−0.654
T-stat	−27.50	−1.047	−31.898

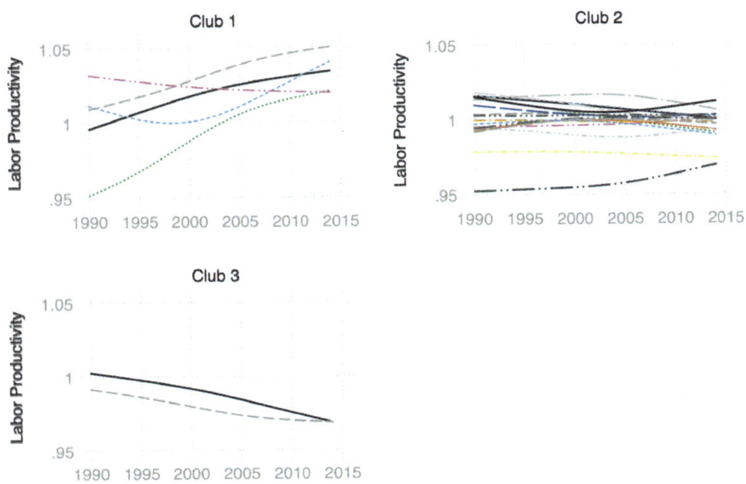

Fig. 4.3 Final clubs and within-club convergence: Labor productivity in developed countries

Club 3, however, is not rejected by the data. Thus, in developed states, the dynamics of labor productivity are characterized by three convergence clubs. Figure 4.3 provides a graphical summary of this final classification (See Appendix A for a list of countries and their respective convergence clubs).

Labor productivity differences across developing countries are larger than those in developed countries, as suggested by Fig. 4.2. Thus, one may expect a larger number of convergence clubs. Table 4.5 helps us evaluate this hypothesis. Initial results indicate that in developing countries labor productivity is characterized by six convergence clubs. The magnitude of the coefficient of the log-t regression suggest that only Club 6 is converging in levels, the rest of the clubs are converging in growth rates. Convergence between clubs is also suggested by the data. Table 4.6 indicates that Club 2 could be merged with either Club 1 or Club 3 (the t-statistic is greater than −1.65 in both cases). Ultimately, Club 2 should be merged with Club 1 as the regression coefficient and t-statistic are larger in this case. Thus, in developing countries, the dynamics of labor productivity appear to be characterized by five convergence clubs.

Based on the transition paths of each individual country in the sample, Fig. 4.4 provides a graphical summary of final club classification. (See Appendix B for a list of countries with their respective club classification.) From this figure, one can clearly observe not only a reduction of productivity disparities within each club, but

Table 4.5 Local convergence test: Labor productivity in developing countries

Statistic	Club1	Club2	Club3	Club4	Club5	Club6
Coeff	0.357	0.291	0.111	0.199	0.124	3.58
T-stat	4.490	4.031	1.517	3.388	0.601	2.26

Table 4.6 Club merge test: Labor productivity in developing countries

Statistic	Club1+2	Club2+3	Club3+4	Club4+5	Club5+6
Coeff	0.101	−0.040	−0.228	−0.212	−0.521
T-stat	1.595	−0.668	−7.926	−4.307	−5.282

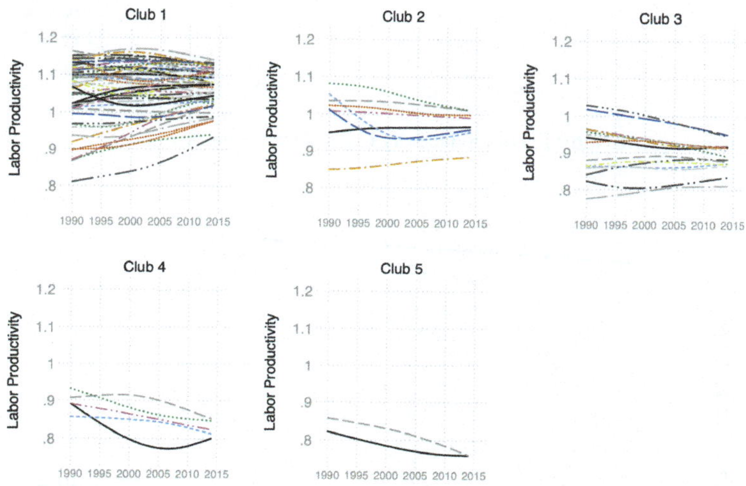

Fig. 4.4 Within-club convergence: Labor productivity in developing countries

also different patterns in terms of the strength of convergence. For instance, at the end of the sample period, productivity differences across countries within Club 1 are far larger than those within Club 5. Overall, this figure also help us understand why Club 5 is converging in levels while the other clubs are only converging in growth rates.

4.3 Transition Paths of the Convergence Clubs

By taking the cross-country average within each club, Figs. 4.5 and 4.5 illustrate the transition paths of all clubs according to the stage of development. These figures are particularly useful to understand the long-run tendencies between clubs. In developed countries (Fig. 4.5), the clubs show clear separating tendencies. On the one hand,

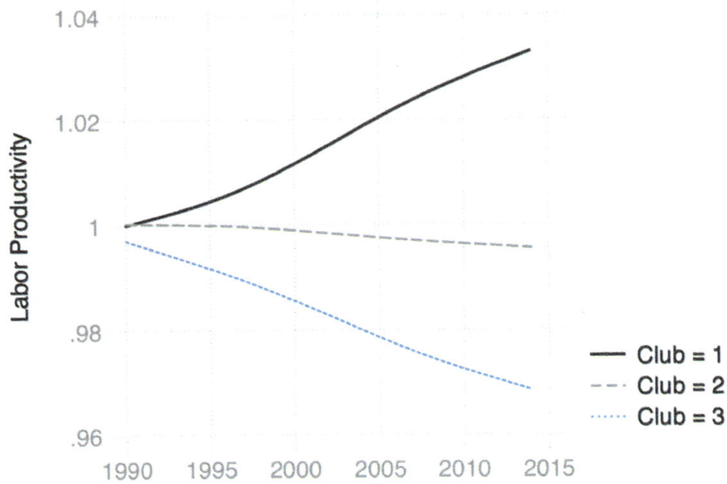

Fig. 4.5 Convergence clubs: Labor productivity in developed countries

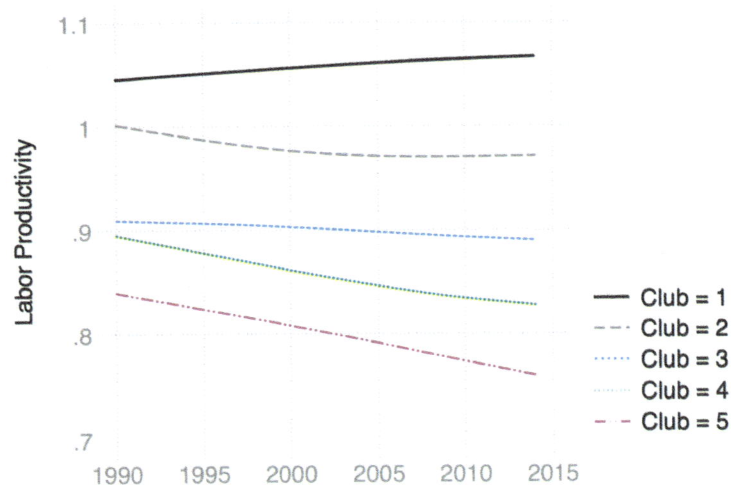

Fig. 4.6 Convergence clubs: Labor productivity in developing countries

Club 1 is systematically diverging from the mean upwards; on the other, Club 3 is diverging from the mean downwards. Although Club 2 has been very close to the mean, it also shows a slight downward divergence pattern. In developing countries (Fig. 4.6), four out of five clubs have been systematically below the mean over the 1990–2014 period. Furthermore, their long-run trends suggest further downward divergence. Club 1, on the other hand, has been diverging from above the mean upwards.

References

Johnson P (2020) Parameter variation in the 'log t' convergence test. Appl Econ Lett 27(9):736–739. https://doi.org/10.1080/13504851.2019.1644436

Phillips P, Sul D (2007) Transition modeling and econometric. Econometrica 75(6):1771–1855

Phillips P, Sul D (2009) Economic transition and growth. J Appl Econ 24(7):1153–1185. https://doi.org/10.1002/jae

Sul D (2019) Panel data econometrics: Common factor analysis for empirical researchers. Routledge

Von Lyncker K, Thoennessen R (2017) Regional club convergence in the EU: evidence from a panel data analysis. Empirical Econ 52(2):525–553. https://doi.org/10.1007/s00181-016-1096-2

Chapter 5
Convergence Clubs in Physical and Human Capital

Abstract This chapter studies convergence in capital accumulation across developed and developing countries, respectively. Through the lens of a non-linear dynamic factor model, the cross-country dynamics of physical and human capital are evaluated over the 1990–2014 period. For developed countries, results reject the hypothesis that all countries would eventually converge to a common steady-state growth path in terms of both physical and human capital. For this group of countries, the dynamics of physical capital (human capital) are characterized by two convergence clubs with largely (weakly) separating trends. For developing countries, results only reject the convergence hypothesis in terms of human capital. The dynamics of physical capital are characterized by a unique convergence club, while the dynamics of human capital are characterized by three largely separated and parallel clubs.

Keywords Club convergence · Physical capital · Human capital · Non-linear dynamic factor model · Developed countries · Developing countries

5.1 Physical Capital Ratio

5.1.1 Testing for Overall Convergence

Figures 5.1 and 5.2 show the transition paths of the capital-output ratio for developed and developing countries, respectively.[1] In these figures, the (physical) capital-output ratio of each country is normalized by the cross-country mean in each year. From a visual inspection, Figs 5.1 and 5.2 suggest patterns of overall convergence. In contrast to the patterns of labor productivity, at the end of the sample period, cross-country differences in the capital-output ratio are smaller than those at the beginning of the sample period.

Tables 5.1 and 5.2 formally evaluate the hypothesis of cross-country convergence based on the log-t convergence test of Phillips and Sul (2007). The log-t test is based

[1] To simplify the presentation of results, the word "capital" in this case only refers to physical capital accumulation.

C. Mendez, *Convergence Clubs in Labor Productivity and its Proximate Sources*, SpringerBriefs in Economics, https://doi.org/10.1007/978-981-15-8629-3_5

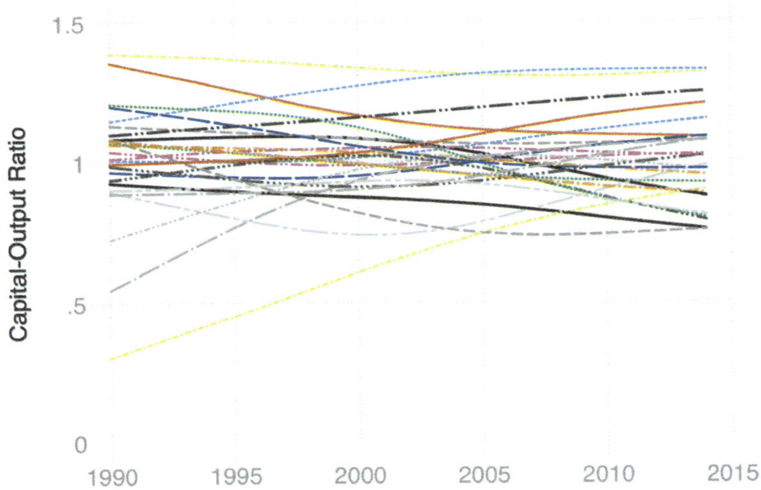

Fig. 5.1 Relative capital-output ratio in developed countries

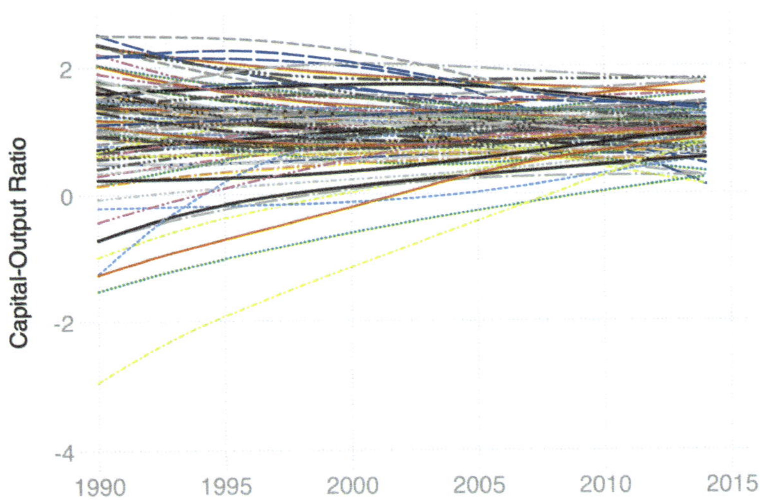

Fig. 5.2 Relative capital-output ratio in developing countries

on a non-linear dynamic factor that allows the modeling of transitional heterogeneity. The results show a very interesting contrast that is not evident by just referring to the figures. The hypothesis of cross-country convergence in the capital-output ratio is rejected by the data of developed countries, but it is not rejected by the data of developing countries. In developing countries, the coefficient of the log-t regression is positive and statistically significant. This result highlights two elements of the convergence process of the capital-output ratio in developing countries. First, cross-country differences have clearly decreased over time. Second, the transition paths

Table 5.1 Global convergence test: Capital-ouput ratio in developed countries

Test	Coeff	SE	T-stat
Log(t)	−0.711	0.051	−13.8

Table 5.2 Global convergence test: Capital-output ratio in developing countries

Test	Coeff	SE	T-stat
Log(t)	0.509	0.054	9.5

of the countries are converging to a common long-run equilibrium. In contrast, in developed countries, the coefficient of the log-regression is negative and statistically significant.

These results also highlight another key feature of the convergence process in developed countries. Although cross-country differences are smaller a the end of the sample period, the transition paths of developed countries are still highly heterogeneous. For instance, from above the mean (Fig. 5.1), some countries clearly show increasing tendencies (positive slopes) while others show decreasing tendencies (negative slopes) in their transition paths. As a result, developed countries may be converging to multiple equilibria in the long run.

5.1.2 *Finding Local Convergence Clubs in Developed Countries*

Although convergence towards a common equilibrium is rejected by the data, the transition paths of developed countries suggest existence of multiple convergence clubs. Table 5.3 formally evaluates this statement by applying the clustering algorithm of Phillips and Sul (2007). Initial results indicate that the capital-output ratio is characterized by three convergence clubs. For each club, the coefficient of the log-t regression is positive and statistically significant. Thus, within each club, physical capital gaps have been declining over time.

The magnitude of the coefficient of the log-t regression is also informative for understanding the strength of the convergence process within each club (Johnson 2020). As this coefficient is less than two in all cases, the data suggest that within clubs countries are converging in terms of their growth rates.

As noted by Phillips and Sul (2009) and Lyncker and Thoennessen (2017), the convergence clubs algorithm of Phillips and Sul (2007) may over state the number of clubs. To handle this concern, it is highly recommended to perform a merge test between the initial clubs. Table 5.4 shows the results of this test. The test suggest that Club 1 and Club 2 can be merged into one enlarged club. Thus, the convergence dynamics of the capital-output ratio in developed countries can be represented by

Table 5.3 Local convergence test: Capital-output ratio in developed countries

Statistic	Club1	Club2	Club3
Coeff	0.635	0.705	0.042
T-stat	7.321	6.565	0.320

Table 5.4 Club merge test: Capital-ouput ratio in developed countries

Statistic	Club1+2	Club2+3
Coeff	0.267	−1.21
T-stat	7.610	−16.65

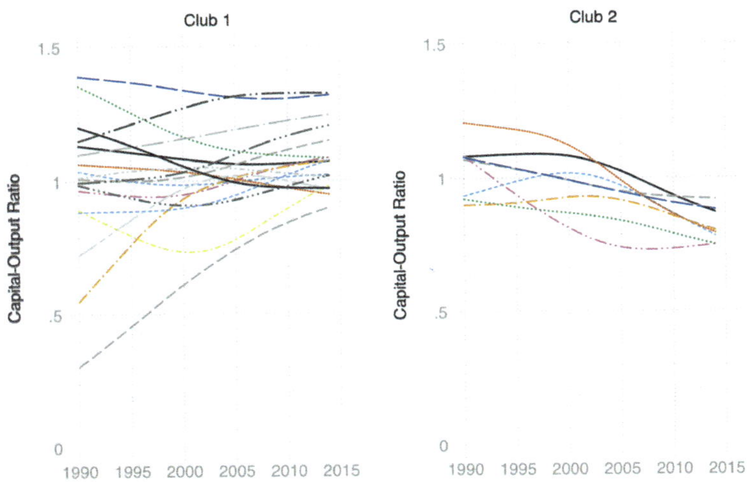

Fig. 5.3 Final clubs and within-club convergence: Capital-output ratio in developed countries

two clubs. Figure 5.3 provides a graphical summary of this final classification (See Appendix A for a list of countries and their respective convergence club).

5.1.3 Transition Paths of the Convergence Clubs in Developed Countries

By taking the cross-country average within each club, Fig. 5.4 illustrates transition paths of clubs in developing nations. This figure is particularly useful to understand the long-run tendencies between clubs. The clubs show clear separating tendencies. On the one hand, Club 1 is systematically diverging from the mean upwards and, on the other, Club 2 is diverging from the mean downwards.

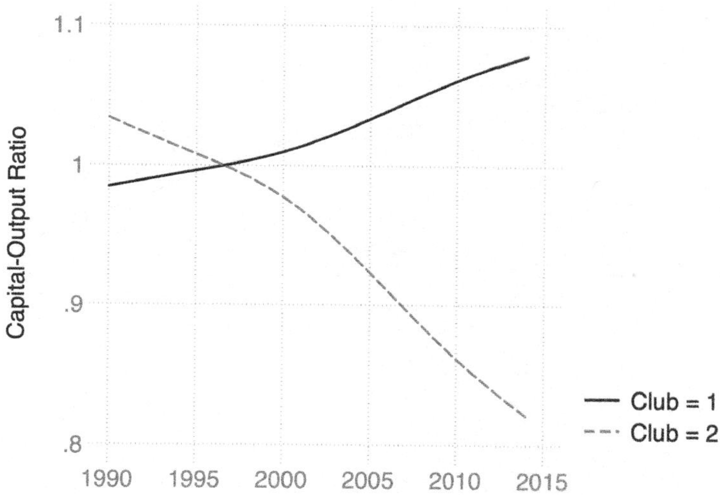

Fig. 5.4 Convergence clubs: Capital-output ratio in developed countries

5.2 Human Capital Per Worker

5.2.1 Testing for Overall Convergence

Figures 5.5 and 5.6 show the transition paths of human capital per worker of developed and developing countries, respectively. From a simple visual inspection, both figures suggest patterns of overall convergence. At the end of the sample period, cross-country disparities in human capital per worker are smaller than those at the beginning of the sample period.

Tables 5.5 and 5.6 formally evaluate the hypothesis of cross-country convergence based on the log-t test of Phillips and Sul (2007). The results contrast the graphical intuition of Figs. 5.5 and 5.6. The hypothesis of cross-country convergence in human capital per worker is rejected by the data of both developed and developing countries. In both groups of countries, the coefficient of the log-t regression is negative and statistically significant.

These results highlight a key element of the cross-country dynamics of human capital. Although cross-country differences are smaller a the end of the sample period, the transition paths of countries are still highly heterogeneous and some countries clearly show diverging tendencies in recent years. As a result, both groups of countries may be converging to multiple equilibria in the long run.

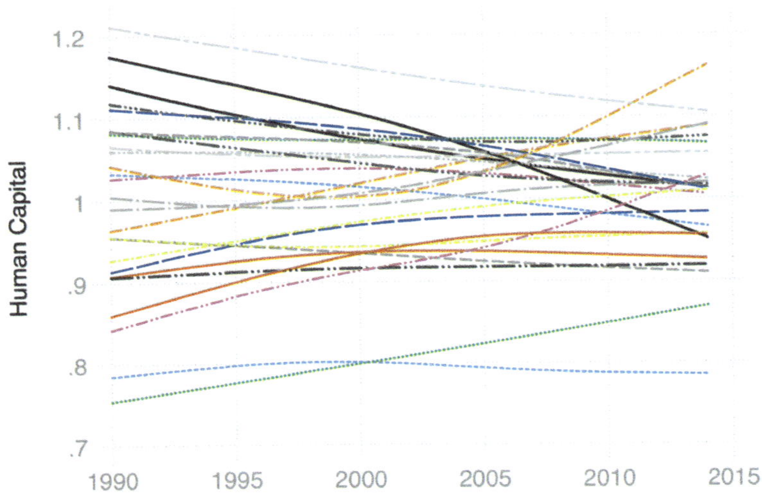

Fig. 5.5 Relative human capital per worker in developed countries

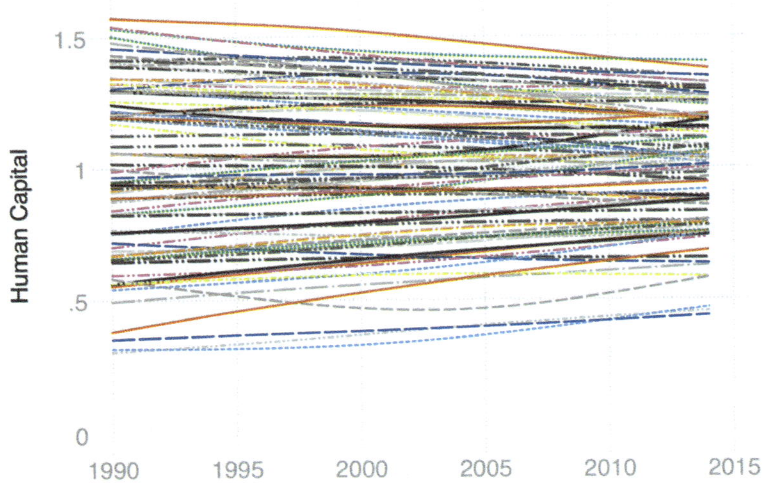

Fig. 5.6 Relative human capital per worker in developing countries

5.2.2 Finding Local Convergence Clubs

Although convergence towards a common equilibrium is rejected by the data, the transition paths of individual countries suggest existence of multiple convergence clubs. Tables 5.7 and 5.8 formally evaluate this hypothesis by applying the clustering algorithm of Phillips and Sul (2007). Initial results indicate that in developed countries, human capital per worker is characterized by two convergence clubs. For

Table 5.5 Global convergence test: Human capital per worker in developed countries

Test	Coeff	SE	T-stat
Log(t)	−0.419	0.039	−10.7

Table 5.6 Global convergence test: Human capital per worker in developing countries

Test	Coeff	SE	T-stat
Log(t)	−0.282	0.034	−8.31

Table 5.7 Local convergence test: Human capital per worker in developed countries

Statistic	Club1	Club2
Coeff	−0.0318	0.0003
T-stat	−0.4414	0.0137

Table 5.8 Local convergence test: Human capital per worker in developing countries

Statistic	Club1	Club2	Club3	Club4	Club5
Coeff	0.287	0.23	0.278	0.091	0.081
T-stat	4.631	3.80	4.208	1.769	1.227

both clubs, the t-statistic is greater than −1.65, thus the hypothesis of within-club convergence cannot be rejected.

Nevertheless, this result may appear inconclusive for two reasons. First, the coefficient of the log-t regression for Club 1 is still negative. Second, the standard error (not reported in Table 5.7) of the coefficient of Club 2 is 0.02, which is larger than the coefficient estimate. In the convergence literature, cases in which the coefficient of the log-t regression and the t-statistic are not strongly consistent are commonly referred as clubs of weak convergence.

In contrast, developing countries show strong patterns of within-club convergence (Table 5.7). Initial results indicate that human capital per worker is characterized by five convergence clubs. For each club, the coefficient of the log-t regression is positive and statistically significant. Thus, within each club, human capital disparities have diminished over time. As previously mentioned, the magnitude of the coefficient of the log-t regression is also informative for understanding the strength of the convergence patterns at the club level. As this coefficient is less than two, all clubs are converging in growth rates.

As previously mentioned, the convergence clubs algorithm of Phillips and Sul (2007) may over-predict the number of clubs. To handle this potential problem, it is highly recommended to perform a sequential merge test between the clubs. Tables 5.9

Table 5.9 Club merge test: Human capital per worker in developed countries

Statistic	Club1+2
Coeff	−0.233
T-stat	−3.997

Table 5.10 Club merge test: Human capital per worker in developing countries

Statistic	Club1+2	Club2+3	Club3+4	Club4+5
Coeff	−0.053	0.197	0.235	−0.11
T-stat	−1.225	3.390	3.863	−2.11

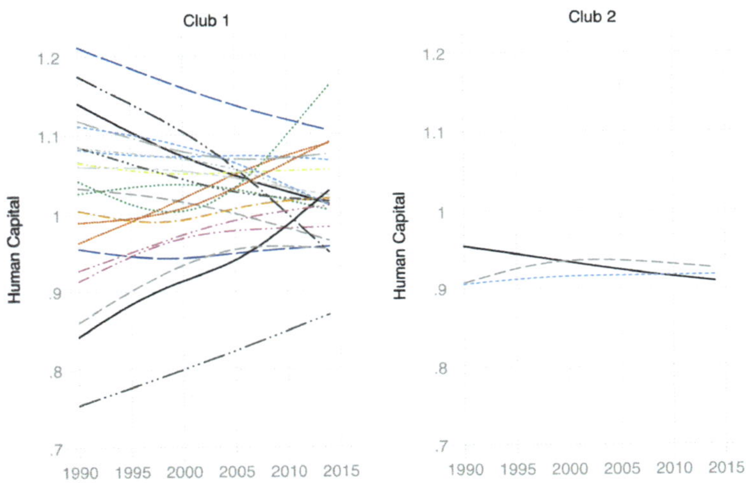

Fig. 5.7 Within-club convergence: Human capital per worker in developed countries

and 5.10 show the results of this test for developed and developing countries, respectively. In developed countries, Club 1 and Club 2 cannot be merged as the regression coefficient is negative and the t-statistic is less than −1.65. In developing countries, however, Club 2, Club 3, and Club 4 can be merged.

Based on the transition paths of each individual country and the results of the merge test, Figs. 5.7 and 5.8 provide a graphical summary of final club classification. (See Appendices A and B for a list of countries with their respective club classification.) In developed countries (Fig. 5.7), the cross-country dynamics of human capital are characterized by two convergence clubs. Most countries belong to Club 1 and only three countries (Austria, Finland, and Italy) belong to Club 2. The case of Portugal (not shown in Fig. 5.7) is particularly interesting as it is the only country that is not converging to any club and its human capital level is declining. In developing countries (Fig. 5.7), the cross-country dynamics of human capital are characterized

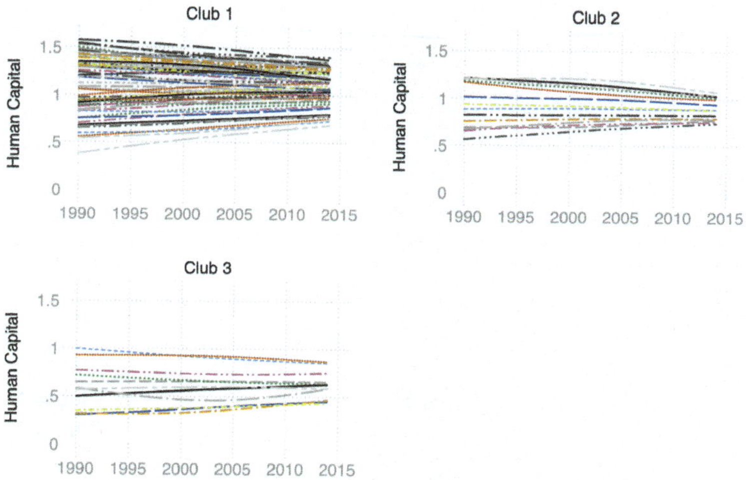

Fig. 5.8 Final clubs and within-club convergence: Human capital per worker in developing countries

by three convergence clubs. Club 1 is the most numerous club, it is composed by 57 countries. Club 2 and Club 3 are composed by 14 and 11 countries, respectively.

5.2.3 Transition Paths of the Convergence Clubs

By taking the cross-country average within each club, Figs. 5.9 and 5.10 illustrate the transition paths at the club level. These figures are particularly useful for understanding long-run tendencies and differences between clubs. In developed countries, Club 1 is systematically above the cross-country mean, which is equal to one, while Club 2 is systematically below it. There are large and increasing differences between the clubs. Although the two clubs showed some small signs of convergence during the 1990–2000 period, since 2000, the gap between the clubs has been widening.

In developing countries, Club 1 is systematically above the cross-country mean, while Club 2 and Club 3 are systematically below it. Over the 1990–2002 period, the lack of convergence among clubs was characterized by three trends that are largely parallel. In the last decade, some progress towards convergence can be observed in Club 3. Relative to the average for developing countries, however, the human capital level of Club 3 is still low an far away from catching up.

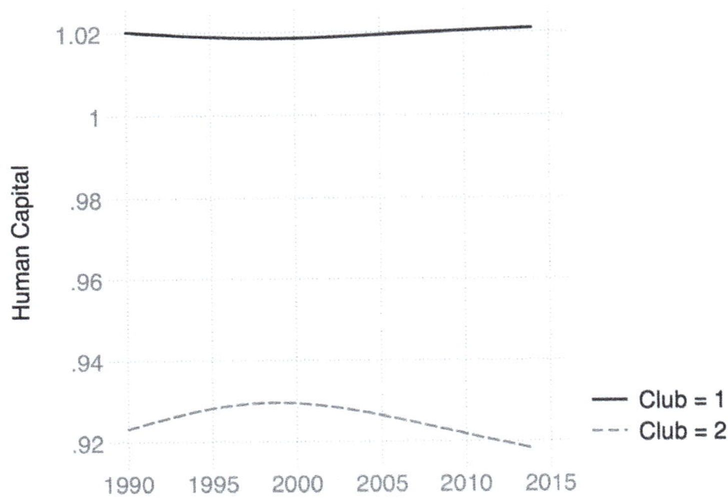

Fig. 5.9 Convergence clubs: Human capital per worker in developed countries

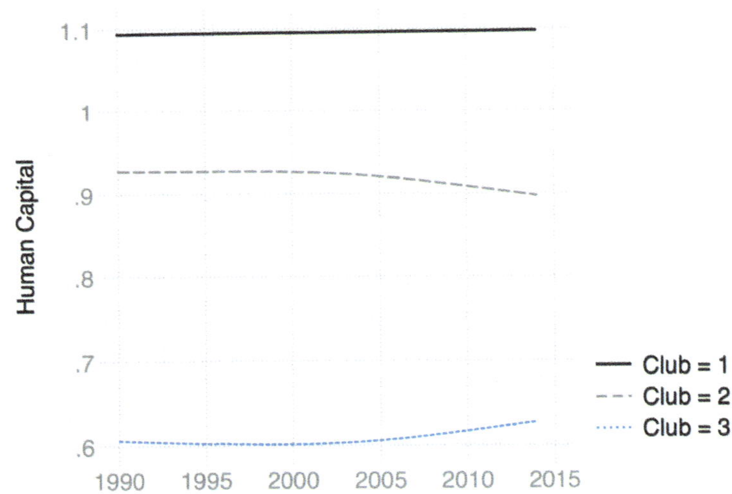

Fig. 5.10 Convergence clubs: Human capital per worker in developing countries

References

Johnson P (2020) Parameter variation in the 'log t' convergence test. Appl Econ Lett 27(9):736–739. https://doi.org/10.1080/13504851.2019.1644436
Phillips P, Sul D (2007) Transition modeling and econometric. Econometrica 75(6):1771–1855

Phillips P, Sul D (2009) Economic transition and growth. J Appl Econ 24(7):1153–1185. https://doi.org/10.1002/jae
Sul D (2019) Panel data econometrics: Common factor analysis for empirical researchers. Routledge
Von Lyncker K, Thoennessen R (2017) Regional club convergence in the EU: evidence from a panel data analysis. Empirical Econ 52(2):525–553. https://doi.org/10.1007/s00181-016-1096-2

Chapter 6
Convergence Clubs in Aggregate Efficiency

Abstract This chapter studies convergence in aggregate efficiency across developed and developing countries, respectively. Through the lens of a non-linear dynamic factor model, the cross-country dynamics of aggregate efficiency are evaluated over the 1990–2014 period. Results reject the hypothesis that all countries would eventually converge to a common steady-state growth path within their respective groups. Developed countries are characterized by three convergence clubs with largely parallel trends. Developing countries are characterized by four convergence clubs with largely widening gaps between the upper and lower tails of the efficiency distribution.

Keywords Club convergence · Aggregate efficiency · Non-linear dynamic factor model · Developed countries · Developing countries

6.1 Testing for Overall Convergence

Figures 6.1 and 6.2 show the transition paths of aggregate efficiency of developed and developing countries, respectively. In these figures, aggregate efficiency of each country is computed as potential (log) GDP per worker divided by a Cobb-Douglass aggregate of the capital inputs (See Chap. 2 for further details). These figures are consistent with the descriptive patterns presented in Chap. 2 in the sense that large efficiency differences across countries are observable during the 1990–2014 period. From a visual inspection, Figs. 6.1 and 6.2 indicate a high degree of efficiency heterogeneity, as well as a lack of overall convergence in developing countries and very little convergence progress in developed countries.

Tables 6.1 and 6.2 formally evaluate the notion of relative convergence based on the log-t convergence test of Phillips and Sul (2007). The log-t test is based on a non-linear dynamic factor that allows the modeling of transitional heterogeneity. As expected, given the patterns of Figs. 6.1 and 6.2, the null hypothesis of convergence in aggregate efficiency is rejected by the data for both groups of countries. For both developed and developing countries, the coefficient of the log-t regression is negative and statistically significant. Thus, cross-country differences in aggregate efficiency have increased over time within both groups of countries.

C. Mendez, *Convergence Clubs in Labor Productivity and its Proximate Sources*, SpringerBriefs in Economics, https://doi.org/10.1007/978-981-15-8629-3_6

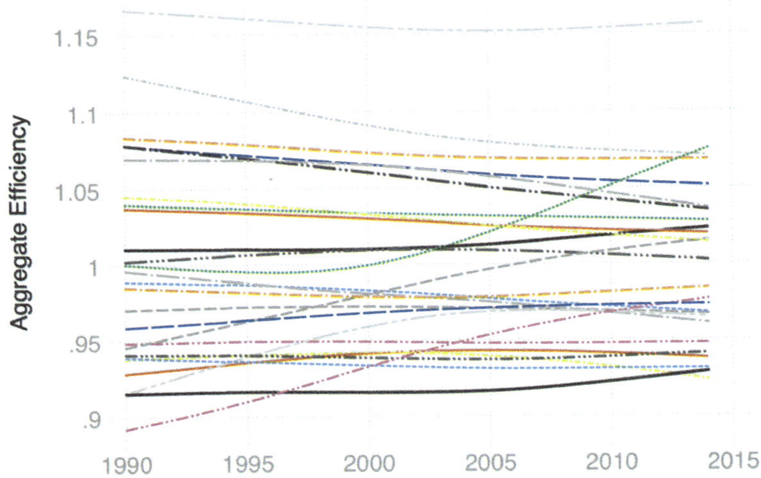

Fig. 6.1 Relative aggregate efficiency in developed countries

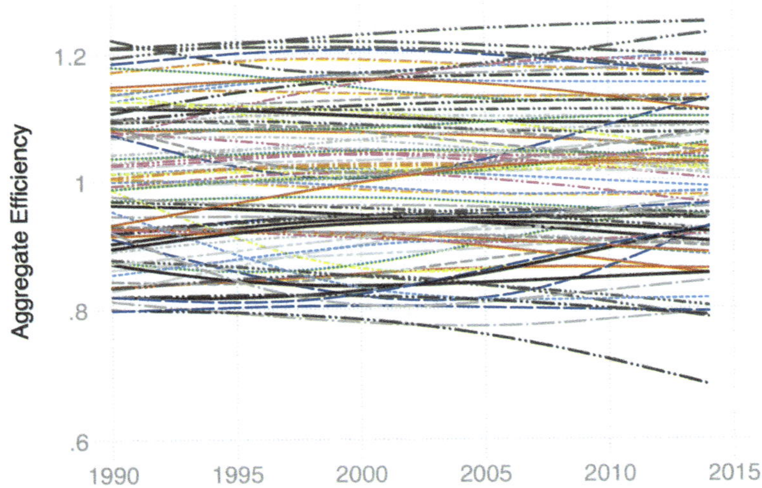

Fig. 6.2 Relative aggregate efficiency in developing countries

Table 6.1 Global convergence test: Aggregate efficiency in developed countries

Test	Coeff	SE	T-stat
Log(t)	−0.569	0.011	−50

Table 6.2 Global convergence test: Aggregate efficiency in developing countries

Test	Coeff	SE	T-stat
Log(t)	−0.712	0.019	−38.5

6.2 Finding Local Convergence Clubs

Although convergence towards a common equilibrium is rejected by the data, the transition paths of individual countries suggest the existence of multiple convergence clubs. Tables 6.3 and 6.4 formally evaluates this intuition by applying the clustering algorithm of Phillips and Sul (2007) to the sample of developed and developing countries, respectively.

In the case of developed countries (Table 6.3), initial results indicate that the dynamics of aggregate efficiency are characterized by three convergence clubs. For Club 1 and Club 2, the coefficient of the log-t regression is positive and statistically significant. Although the coefficient of Club 3 is negative, the t-statistic is greater than −1.65. Thus, it could be described as a club of weak convergence. The magnitude of the coefficient of the log-t regression is also informative for understanding the strength of the convergence process within each club (Johnson 2020). As this coefficient is less than two for all cases (Table 6.3), the clubs of developed countries are converging in terms of growth rates.

In the case of developing countries (Table 6.4), initial results indicate that the dynamics of aggregate efficiency are characterized by six convergence clubs. With the exception of Club 4, the coefficient of the log-t regression is positive and statistically significant in all clubs. Although the coefficient of Club 4 is negative, the t-statistic is greater than −1.65. Thus, it could be described as a club of weak convergence. The magnitude of the coefficient of the log-t regression gives further information regarding the strength of the convergence process within each club (Johnson 2020). As this coefficient is less than two for all cases in Table 6.4, it suggests that the clubs of developing countries are converging in terms of growth rates.

As noted by Phillips and Sul (2009) and Lyncker and Thoennessen (2017), the convergence clubs algorithm of Phillips and Sul (2007) may over-estimate the number of clubs. To handle this potential issue, it is highly recommended to run a sequential merge test between the clubs.

For developed countries, Table 6.5 shows the results of the merge test for each sequential club. As the regression coefficient is negative and the t-statistic is less

Table 6.3 Local convergence test: Aggregate efficiency in developed countries

Statistic	Club1	Club2	Club3
Coeff	0.093	0.05	−0.008
T-stat	3.536	3.08	−0.067

Table 6.4 Local convergence test: Aggregate efficiency in developing countries

Statistic	Club1	Club2	Club3	Club4	Club5	Club6
Coeff	0.487	0.136	0.236	−0.106	0.577	0.434
T-stat	3.239	1.711	3.520	−1.608	3.994	3.663

Table 6.5 Club merge test: Aggregate efficiency in developed countries

Statistic	Club1+2	Club2+3
Coeff	−0.427	−0.319
T-stat	−19.497	−16.609

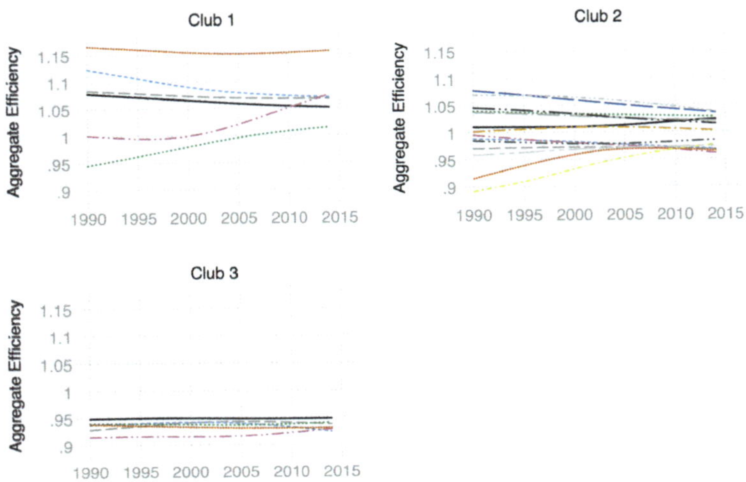

Fig. 6.3 Final clubs and within-club convergence: Aggregate efficiency in developed countries

Table 6.6 Club merge test: Aggregate efficiency in developing countries

Statistic	Club1+2	Club2+3	Club3+4	Club4+5	Club5+6
Coeff	0.055	0.135	−0.233	−0.306	−0.443
T-stat	0.700	1.893	−4.712	−5.952	−6.622

than −1.65, the hypothesis of convergence between clubs is rejected by the data. Thus, the dynamics of aggregate efficiency in developed countries are ultimately characterized by three convergence clubs. Figure 6.3 provides a graphical summary of this final classification (See Appendix A for a list of countries and their respective convergence club).

For developing countries, Table 6.6 shows the results of the merge test. As the regression coefficient is positive and the t-statistic is greater than -1.65, the hypothesis of convergence between clubs is not rejected for Club 1, Club 2, and Club 3.

These three clubs can be merged into one larger club. After this merge, the dynamics of aggregate efficiency in developing countries are ultimately characterized by four convergence clubs. Figure 6.4 provides a graphical summary of this final classification (See Appendix B for a list of countries and their respective convergence club).

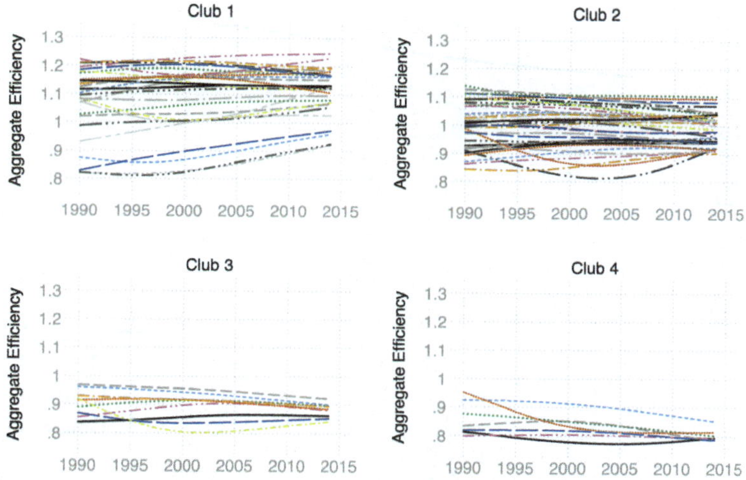

Fig. 6.4 Within-club convergence: Aggregate efficiency in developing countries

6.3 Transition Paths of the Convergence Clubs

By taking the cross-country average within each club, Figs. 6.5 and 6.6 illustrate the transition paths at the club level for developed and developing countries, respectively. In developed countries (Fig. 6.5), the long-run tendencies of the three clubs do not show signs of convergence. Club 1 is systematically above the cross-country mean, which is equal to one and, in the last decade, it has tended to separate further from

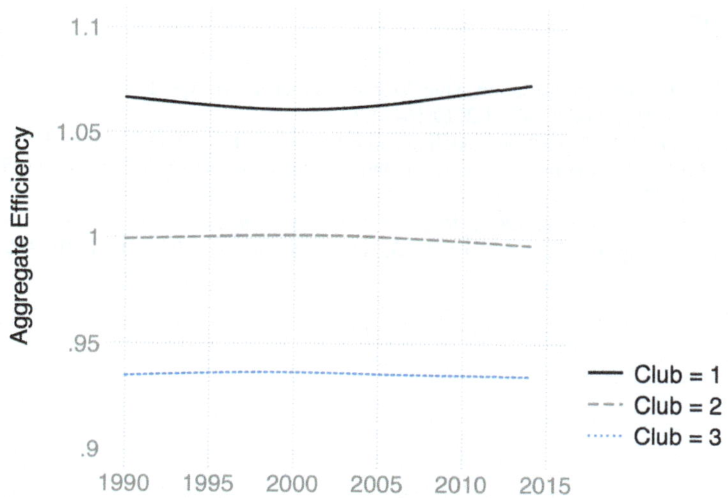

Fig. 6.5 Convergence clubs: Aggregate efficiency in developed countries

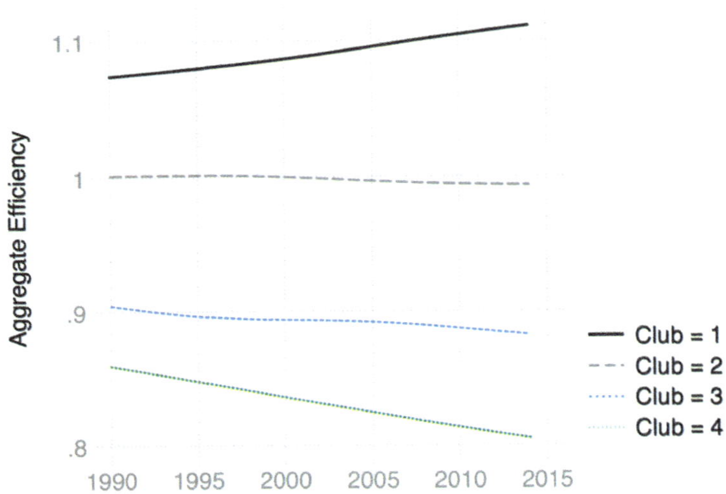

Fig. 6.6 Convergence clubs: Aggregate efficiency in developing countries

the mean. Club 3, on the other hand, is systematically below the mean and, over the entire sample period, it has shown a flat tendency. Club 2 is located in the vicinity of the cross-country mean. In developing countries (Fig. 6.6), the long-run tendencies of the four clubs do not show signs of convergence. Moreover, there are clear separating tendencies between Club 1 and Club 4. Club 2 and Club 3 show slight downward tendencies.

References

Johnson P (2020) Parameter variation in the 'log t' convergence test. Appl Econ Lett 27(9):736–739. https://doi.org/10.1080/13504851.2019.1644436

Phillips P, Sul D (2007) Transition modeling and econometric. Econometrica 75(6):1771–1855

Phillips P, Sul D (2009) Economic transition and growth. J Appl Econ 24(7):1153–1185. https://doi.org/10.1002/jae

Von Lyncker K, Thoennessen R (2017) Regional club convergence in the EU: evidence from a panel data analysis. Empirical Econ 52(2):525–553. https://doi.org/10.1007/s00181-016-1096-2

Chapter 7
Concluding Remarks

Abstract This chapter summarizes the main takeaways of the book. It first outlines the main empirical findings for developed and developing countries, respectively. These findings suggest that the convergence clubs of aggregate efficiency are more likely to explain the dynamics of the clubs of labor productivity. Next, it lists the methodological takeaways from the convergence framework of Phillips and Sul (2007a, b) and (2009). Finally, it points out some emerging research directions and methodological advances.

Keywords Club convergence · Labor productivity · Capital accumulation · Aggregate efficiency · Developed countries · Developing countries

7.1 Summing Up All Findings

In developed countries:

- Labor productivity is characterized by three convergence clubs with largely separating trends.
- Both physical capital and human capital are characterized by two convergence clubs. Relative to human capital, the club trends of physical capital show stronger divergence patterns.
- Aggregate efficiency is characterized by three convergence clubs with largely parallel trends.

In developing countries:

- Labor productivity is characterized by five convergence clubs in which the bottom clubs show trends largely below the mean and diverging downwards.
- Physical capital and human capital show opposite dynamics. Physical capital is characterized by a unique convergence club, while human capital is characterized by three largely separating clubs.
- Aggregate efficiency is characterized by four convergence clubs with largely separating trends between the upper and lower tails of the distribution.

C. Mendez, *Convergence Clubs in Labor Productivity and its Proximate Sources*, SpringerBriefs in Economics, https://doi.org/10.1007/978-981-15-8629-3_7

The existence of multiple convergence clubs in labor productivity highlights the well documented fact that there are large productivity differences across countries. As expected, larger productivity differences are observed across developing countries. The club dynamics of the capital accumulation and aggregate efficiency provide a first explanation regarding the sources of those productivity differences. Relative to the convergence clubs of capital accumulation, convergence club differences in aggregate efficiency are larger and more predominant. Thus, it seems that most of the club convergence differences in labor productivity are explained by club convergence differences in aggregate efficiency.

7.2 Some Methodological Takeaways

- The modern convergence framework of Phillips and Sul (2007a, b), and Phillips and Sul (2009) is based on the role of technological heterogeneity and multiple equilibria.
- The transitional path of an economy i at time t can be modeled by $y_{i,t} = \delta_{it}\mu_t$, where μ_t is a common long-run growth path and δ_{it} an idiosyncratic transition path.
- Convergence across economies is defined as $\lim_{t\to\infty} h_{it} = 1$, where $h_{it} = ny_{it}/\sum_{j=1}^{n} y_{jt} = n\delta_{it}/\sum_{j=1}^{n}\delta_{jt}$ and n is the number of economies.
- To empirically test for convergence, Phillips & Sul (2007b) proposed the following (log-t) regression model:

$$\log\frac{H_1}{H_t} - 2\log(\log t) = a + b\log t + u_t,$$

  where $H_t = \frac{1}{n}\sum_{i=1}^{n}(h_{it} - 1)^2$ is variance of h_{it}. Based on this model, a one-sided t-test is used to evaluate the convergence hypothesis: $b > 0$.
- The estimate of the parameter b provides additional information about the convergence process: $b < 0$ implies divergence, $0 < b < 2$ implies convergence in growth rates, and $b \geq 2$ implies convergence in levels.
- When the convergence hypothesis is rejected ($b < 0$), a clustering algorithm for panel data is implemented to identify local convergence clubs.

7.3 New Research Directions

The analyses of relative convergence and club convergence are rapidly evolving toward multiple research directions. Three recent methodological developments are particularly noteworthy. First, although many recent studies have evaluated the determinants of the convergence clubs, spatial ordered logit models provide a new framework to evaluate the role of space in the formation of clubs (Li et al. 2018). Second, relative convergence can now be evaluated even when the data do not involve strong deterministic or stochastic trends (Kong et al. 2019). Third, locally linear versions of the log-t test of convergence are particularly useful to evaluate how the strength of convergence changes over time (Johnson 2020).

Since the pioneer work of Bartkowska and Riedl (2012), the analysis of club convergence is typically implemented in two stages. In the first stage, the convergence clubs are identified using the test and clustering algorithm of Phillips and Sul (2007b). In the second stage, the determinants of the clubs are evaluated through the lens of an ordered logit model. Variables such as initial income, capital accumulation, and sectorial output shares are commonly reported as statistically significant predictors of the convergence clubs. The study of Li et al. (2018) extends this second stage by implementing a dynamic spatial ordered probit model. This new research direction is particularly useful to evaluate the role of spatial dependence on the formation and dynamics of the convergence clubs.

In the original framework of Phillips and Sul (2007b) and Phillips and Sul (2009), the analyses of relative convergence and club convergence are limited to time series that exhibit clear and common (stochastic or deterministic) long-run trends. However, some important economic variables, such as unemployment, do not show strong trends and are weakly dependent. To handle this type of variables, Kong et al. (2019) have developed the notion of weak sigma-convergence. Based on this approach, it is now possible to study convergence in asymptotically stationary or weakly dependent series.

The original framework of Phillips and Sul (2007a, b), and Phillips and Sul (2009) mostly focuses on the role of technological heterogeneity across economic units. The study of heterogeneity over time, however, is limited by the linearized specification of the log-t convergence test. To handle this limitation, Johnson (2020) proposes a locally linear version of the log-t regression model. This methodological extension is particularly useful to evaluate how the forces of convergence change over time.

References

Bartkowska M, Riedl A (2012) Regional convergence clubs in europe: identification and conditioning factors. Econ Model 29(1):22–31. https://doi.org/10.1016/j.econmod.2011.01.013

Johnson P (2020) Parameter variation in the 'log t' convergence test. Appl Econ Lett 27(9):736–739. https://doi.org/10.1080/13504851.2019.1644436

Kong J, Phillips PCB, Sul D (2019) Weak-convergence: Theory and applications. J Econ 209(2):185–207. https://doi.org/10.1016/j.jeconom.2018.12.022

Li F, Li G, Qin W, Qin J, Ma H (2018) Identifying economic growth convergence clubs and their influencing factors in China. Sustainability (Switzerland) 10(8):1–21. https://doi.org/10.3390/su10082588

Phillips P, Sul D (2007a) Some empirics on economic growth under heterogeneous technology. J Macroecon 29(3)455–469. https://doi.org/10.1016/j.jmacro.2007.03.002

Phillips P, Sul D (2007b) Transition modeling and econometric. Econometrica 75(6):1771–1855

Phillips P, Sul D (2009) Economic transition and growth. J Appl Econ 24(7):1153–1185. https://doi.org/10.1002/jae

Appendix A
List of Developed Countries and Their Convergence Clubs

Country	LP	KY	HC	AE
Australia	2	2	1	2
Austria	2	1	2	2
Belgium	2	1	1	2
Canada	2	2	1	2
Denmark	2	1	1	3
Finland	2	1	2	3
France	2	1	1	1
Germany	2	1	1	1
Greece	2	1	1	3
Hong Kong	2	1	1	2
Ireland	1	1	1	2
Israel	3	2	1	3
Italy	2	1	2	2
Japan	NA	1	1	1
Netherlands	2	1	1	2
New Zealand	3	2	1	3
Norway	1	2	1	1
Portugal	2	1	NA	3
Saudi Arabia	1	2	1	1
Singapore	1	1	1	2
Spain	2	1	1	2
Sweden	2	1	1	2
Switzerland	2	2	1	2
Taiwan	2	1	1	2
United Kingdom	2	1	1	2
United States	1	2	1	1

Note NA indicates a country that is not converging to any club. LP stands for labor productivity, KY stands for physical capital-to-output ratio, HC stands for human capital per worker, and AE stands for aggregate efficiency.

C. Mendez, *Convergence Clubs in Labor Productivity and its Proximate Sources*, SpringerBriefs in Economics, https://doi.org/10.1007/978-981-15-8629-3

Appendix B
List of Developing Countries and Their Convergence Clubs

Country	LP	KY	HC	AE
Albania	1	1	1	2
Algeria	1	1	1	1
Argentina	1	1	1	1
Armenia	1	1	1	1
Bangladesh	3	1	1	2
Benin	3	1	1	3
Bolivia	2	1	2	2
Brazil	1	1	1	1
Bulgaria	1	1	1	2
Burundi	5	1	3	4
Cambodia	3	1	2	2
Cameroon	3	1	2	3
Central African Republic	5	1	3	NA
Chile	1	1	1	2
China	1	1	1	1
Colombia	1	1	1	2
Congo	1	1	3	2
Costa Rica	1	1	2	2
Cote d'Ivoire	3	1	2	2
Croatia	1	1	1	2
Czech Republic	1	1	1	2
Democratic Republic of Congo	4	1	3	2
Dominican Republic	1	1	1	2
Ecuador	1	1	2	2
Egypt	1	1	1	1
El Salvador	1	1	1	1

(continued)

C. Mendez, *Convergence Clubs in Labor Productivity and its Proximate Sources*, SpringerBriefs in Economics, https://doi.org/10.1007/978-981-15-8629-3

(continued)

Country	LP	KY	HC	AE
Estonia	1	1	1	2
Ghana	3	1	2	2
Guatemala	1	1	3	2
Haiti	4	1	2	4
Honduras	3	1	2	3
Hungary	1	1	1	2
India	1	1	1	1
Indonesia	1	1	1	1
Iran	1	1	1	1
Jamaica	2	1	1	4
Kazakhstan	1	1	1	1
Kenya	3	1	3	2
Kyrgyz Republic	2	1	1	2
Latvia	1	1	1	2
Lithuania	1	1	1	2
Malawi	4	1	1	4
Malaysia	1	1	1	1
Mali	3	1	3	3
Mexico	1	1	1	1
Mongolia	1	1	1	1
Morocco	2	1	1	2
Mozambique	3	1	3	3
Myanmar	1	1	1	1
Nepal	3	1	1	2
Nicaragua	3	1	1	3
Niger	NA	1	3	4
Pakistan	2	1	2	1
Panama	1	1	1	2
Paraguay	1	1	1	2
Peru	1	1	2	1
Philippines	2	1	1	2
Poland	1	1	1	1
Romania	1	1	1	1
Russia	1	1	1	1
Rwanda	3	1	2	3
Senegal	3	1	3	3
Sierra Leone	4	1	1	4
Slovak Republic	1	1	1	2
Slovenia	1	1	1	2
South Africa	1	1	1	1
South Korea	1	1	1	1
Sri Lanka	1	1	1	2
Sudan	1	1	3	2
Tajikistan	2	1	1	3
Tanzania	2	1	1	2
Thailand	1	1	1	2

(continued)

(continued)

Country	LP	KY	HC	AE
Togo	4	1	2	4
Tunisia	1	1	1	2
Turkey	1	1	1	1
Uganda	3	1	1	2
Ukraine	1	1	1	1
Uruguay	1	1	2	2
Venezuela	1	1	1	1
Vietnam	1	1	1	2
Yemen	1	1	1	1
Zambia	1	1	1	1

Note NA indicates a country that is not converging to any club. LP stands for labor productivity, KY stands for physical capital-to-output ratio, HC stands for human capital per worker, and AE stands for aggregate efficiency.

The manufacturer's authorised representative in the EU is Springer Nature Customer Service Centre GmbH, Europaplatz 3, 69115 Heidelberg, Germany. If you have any concerns regarding our products, please contact ProductSafety@springernature.com

Printed and bound by CPI Group (UK) Ltd, Croydon, CR0 4YY
30/09/2026
02225261-0004